Automatic Tuning of PID Controllers

Karl Johan Åström and Tore Hägglund

INSTRUMENT SOCIETY OF AMERICA

**Automatic Tuning
of PID Controllers**

Copyright ® Instrument Society of America 1988

All rights reserved

Printed in the United States of America

No part of this publication may be reproduced, stored in a retrieval system, or transmitted, in any form or by any means, electronic, mechanical, photocopying, recording or otherwise, without the prior written permission of the publisher.

INSTRUMENT SOCIETY OF AMERICA
67 Alexander Drive
P.O. Box 12277
Research Triangle Park, NC 27709

Library of Congress Cataloging-in-Publication Data

Aström, Karl J. (Karl Johan), 1934
    Automatic tuning of PID controllers.

    Bibliography: p.
    1. PID controllers. I. Hägglund, Tore. II. Instrument Society of America. III. Title.

TJ223.P55A87 1988   629.8   88-3010
ISBN 1-55617-081-5

Book design by Summit Technical Associates, Inc.

Preface

Having grown out of the authors' research, this text summarizes the state of the art of PID control and methods for automatic tuning of PID controllers. The work originated as a small project to tune a simple regulator. As the work progressed we learned more and more about the problem and the industrial use of PID control. We came across many useful ideas used by practitioners of control but rarely mentioned in the standard control courses. We also became increasingly convinced of the usefulness of automatic tuning devices. This was reinforced strongly by plant experiments and the feedback obtained from instrument engineers. We also found that there were many challenging problems in the field. Finally, we became convinced that a new generation of PID controllers is emerging and developing. It has been a pleasure and a privilege to participate in this development firsthand.

We would like to express our gratitude to several persons and agencies who have provided support and inspiration. Axel Westrenius of Telemetric contributed much knowledge of control practices, and Mike Sommerville of Eurotherm shared his ideas about practical PID control. We have benefitted from discussions with Edgar Bristol and Greg Shinskey of Foxboro and Ken Goff of Leeds and Northrup. Particular thanks are due to Sune Larsson of Satt Control who got us interested in industrial uses of autotuning and Lars Bååth of the same company with whom we shared the pleasures and perils of developing an industrial autotuner. Many thanks are due to our colleagues at the Department of Automatic Control for useful discussions, in particular Per-Olof Olsson and Anders Wallenborg who read an early version of the manuscript and Professor C. C. Hang at Singapore University who read a late version. We are grateful to many instrument engineers who participated in experiments and who generously shared their experiences with us. We are very grateful to Agneta Tuszynski who meticulously typed several versions of the manuscript. We would also like to thank the Swedish Board of Technical development (STU) who generously have supported our research.

Table of Contents

PREFACE iii
1 INTRODUCTION 1
2 PID CONTROL 3
2.1 Introduction 3
2.2 The Basic Algorithm 4
2.3 Integral Windup 10
2.4 Reference Values 14
2.5 Digital Implementation 16
2.6 When Can PID Control Be Used? 24
2.7 Conclusions 28
3 PROCESS DYNAMICS 29
3.1 Introduction 29
3.2 Transient Response 29
3.3 Frequency Response 37
3.4 Parameter Estimation 45
3.5 Conclusions 50
4 DESIGN OF PID CONTROLLERS 51
4.1 Introduction 51
4.2 Ziegler-Nichols Methods 52
4.3 Dominant Pole Design 62
4.4 Frequency Domain Design 74
4.5 Pole Placement 81
4.6 Discrete Time Pole Placement 88
4.7 Improvement of Set Point Control 92
4.8 Comparison of the Design Methods 95
4.9 Conclusions 103
5 AUTOTUNING 105
5.1 Introduction 105
5.2 Approaches to Autotuning 106
5.3 The Foxboro Exact™ 110
5.4 The Satt Control Instruments Autotuner™ 116
5.5 The Leeds & Northrup Electromax V™ 120
5.6 The Turnbull Control Systems 6355
       Autotuning Controller™ 126
5.7 Conclusions 132
6 CONCLUSIONS 133
REFERENCES 137
ABOUT THE AUTHORS 143

Introduction

1

PID controllers, the bread and butter of control engineering practice, are found in large numbers in all industries. They come in many different forms, are packaged as standard products, and are manufactured by the hundred thousands yearly. PID controllers are also embedded in all kinds of special purpose control systems. They have survived many changes in technology ranging from pneumatics to electron tubes, transistors, integrated circuits, and microprocessors.

These controllers have several important functions: they provide feedback, they have the ability to eliminate steady-state offsets through the integral action, they can anticipate the future through the derivative action, and they can cope with actuator saturation. Much good control practice is engineered into them. PID controllers are also sufficient for many control problems, particularly where there are benign process dynamics and modest performance requirements. They are thus important components in the control engineer's toolbox. Together with logic, sequential machines, selectors, and simple function blocks, they are used to build the automation equipment for energy production, transportation, and manufacturing, which is an important part of today's technology.

A large cadre of instrument and process engineers are familiar with the operation of PID controllers. There is also a well-established practice of installing, tuning, and using them.

The microprocessor has had a dramatic effect on PID controllers as on other types of industrial electronics, and a large number of those manufac-

Introduction

tured today are based on microprocessors. The microprocessor-based systems were initially pure translations of previous technology. The fundamental work in understanding how to do this was performed in the late sixties and early seventies in connection with the development of direct digital control systems on minicomputers. It should be kept in mind that the computing power of today's micros often far exceeds the power of those minis. As the potentials of the micros are being realized by manufacturers, new features continually appear.

Although PID controllers are common and well-known, they are often poorly tuned. Evidence for this can be found in the control rooms of any industry. The derivative action is frequently switched off for the simple reason that it is difficult to tune properly. It is no coincidence that the derivative action can be switched off in most of the controllers that provide this function.

The microprocessor offers interesting possibilities to provide automatic tuning as well as adaptation to slowly changing operating conditions. The terminology in these areas is not well-established. For purposes of this text, *autotuning* means that the controller parameters are tuned automatically on demand from an operator or an external signal, and *adaptation* means that the parameters of a controller are continuously updated. Commercial products with these capabilities are starting to appear on the market.

The authors believe that a new era of PID control is emerging. We would like to take stock of the development, assess its potential, and try to speed up the development by sharing our experiences in this exciting application of automatic control. The attention is focused on automatic tuning, but adaptation will also be discussed briefly.

The PID controller is discussed in depth in Chapter 2. This includes principles as well as many implementation details such as limitation of derivative gain, anti-windup, improvement of set point response, etc. Different methods of obtaining process models are discussed in Chapter 3. Chapter 4 describes methods for the design of PID controllers. In Chapter 5 the ideas from Chapters 3 and 4 are combined to obtain autotuning devices. The principles are first discussed and a number of commercial autotuners are then described. Some conclusions are summarized in Chapter 6, where an attempt is also made to evaluate the different approaches with respect to the industrial needs. A short discussion of adaptive control and expert control is also given.

It is assumed that the reader has a control background. Even so, the explanations are elementary. Occasionally, we have stated facts without supporting detailed arguments when they have seemed unnecessary, in an effort to focus on the practical aspects rather than the theory.

PID Control

2

2.1 INTRODUCTION

The PID controller is by far the most common control algorithm. Most feedback loops are controlled by this algorithm or minor variations of it. It is implemented in many different forms, as a stand-alone regulator or as a part of a DDC package or hierarchical distributed process control system. Many thousands of instrument and control engineers worldwide are using such regulators in their daily work. The PID algorithm can be approached from many different angles. It can be viewed as a device that can be operated with a few rules of thumb, but it can also be approached analytically.

This chapter gives a short introduction to PID control. The basic algorithm is presented as well as a description of the properties of the controller in a closed-loop based on intuitive arguments and representations of the controller. The phenomenon of reset windup, which occurs when a regulator with integral action is connected to a process with a saturating actuator, is discussed, including methods to avoid it. Different ways of introducing the reference value into the feedback loop are presented, leading to a treatment of feedforward. Some important aspects of digital computer implementation of PID controllers are given: issues such as prefiltering, different digital approximations, noise filtering and the computer code for good implementation. Some aspects on the use and misuse of PID control are presented with examples of systems where PID control works well and where it does not.

PID Control

2.2 THE BASIC ALGORITHM

The "textbook" version of the PID algorithm has the following form:

$$u(t) = K\left[e(t) + \frac{1}{T_i}\int e(s)\,ds + T_d\frac{de(t)}{dt}\right] \qquad (2.1)$$

where u is the control variable and e is the control error ($e = r - y$), which is the difference between set point r and measured value y. The control variable is thus a sum of three terms: the P-term (which is proportional to the error), the I-term (which is proportional to the integral of the error), and the D-term (which is proportional to the derivative of the error). The controller parameters are proportional gain K, integral time T_i, and derivative time T_d.

Proportional Action

In the case of pure proportional control, the control law of Equation 2.1 reduces to

$$u(t) = Ke(t) \qquad (2.2)$$

The control action is simply proportional to the control error. This is the simplest form of feedback. Several properties of proportional control can be understood by the following argument, which is based on pure static considerations. Consider the simple feedback loop, shown in Figure 2.1 and composed of a process and a controller. Assume that the controller has proportional action and that the process is modeled by the static model

$$x = K_p u \qquad (2.3)$$

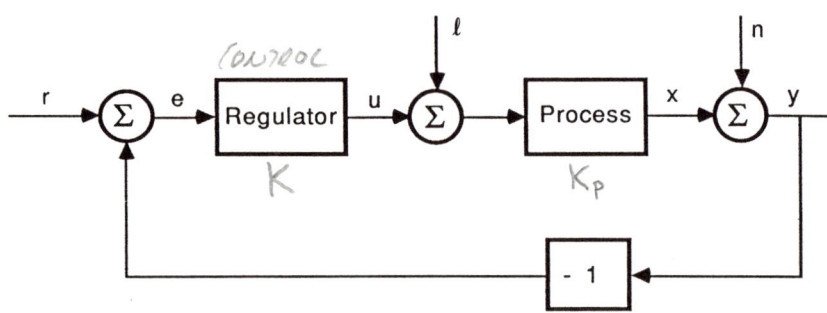

Figure 2.1
Block Diagram of a Simple Feedback Loop

PID Control

The following equations are obtained from the block diagram.

$$y = x + n$$
$$x = K_p(u + \ell)$$
$$u = K(r - y)$$

Elimination of intermediate variables gives the following relation between process variable x, set point r, load disturbance ℓ, and measurement noise n:

$$x = \frac{KK_p}{1 + KK_p}(r - n) + \frac{K_p}{1 + KK_p} \ell \qquad (2.4)$$

Product KK_p is a dimensionless number called the *loop gain*. Several interesting properties of the closed-loop system can be read from Equation 2.4. The loop gain should be high in order to ensure that process output x is close to set point r. A high value of the loop gain will also make the system insensitive to load disturbance ℓ. It follows from Equation 2.4 that measurement noise n influences the process output in the same way as set point r. A high loop gain thus makes the system sensitive to measurement noise.

It also follows from Equation 2.4 that there will always be a steady-state error with proportional control. This can be deduced intuitively from the observation that it follows from Equation 2.2 that a control error is necessary in order to have a non-zero control signal. Proportional controllers are therefore often provided with a reset term I to obtain a correct steady state. The control law (Equation 2.2) then becomes

$$u(t) = K\,e(t) + I \qquad (2.5)$$

where I is the reset term. The above arguments, which are based on the assumption that the process can be described by a static model, leaves out some important properties of the closed-loop system dynamics. The most important one is that the closed-loop system will normally be unstable for high loop gains if the process dynamics are considered. In practice, the maximum loop gain is thus determined by the process dynamics.

One way to describe process dynamics leads to descriptions like Equation 2.4, where the process gain is frequency-dependent.

Integral Action

The main function of the integral action is to make sure that the process output agrees with the set point in steady state. With proportional control, it

PID Control

is necessary to have an error in order to have a non-zero control signal. With integral action, a small positive error will always lead to an increasing control signal, and a negative error will give a decreasing control signal no matter how small the error is.

The following simple argument shows that the steady-state error will always be zero. Assume that the system is in steady state with a constant control signal (u_0) and a constant error (e_0). It follows from Equation 2.1 that the control signal is then given by

$$u_0 = K \left[e_0 + \frac{e_0}{T_i} t \right]$$

As long as $e_0 \neq 0$, this clearly contradicts the assumption that u_0 is constant. A controller with integral action will thus always give zero steady-state error.

Integral action can also be motivated as a device that automatically adjusts the reset of a proportional controller. This is illustrated in the block diagram in Figure 2.2, which shows a proportional controller with a reset that is adjusted automatically. The adjustment is made by feeding back a signal, which is a filtered value of the output, to the summing point of the controller. This was actually one of the early inventions of integral action or "automatic reset" as it was also called.

The implementation shown in Figure 2.2 is still used by many manufacturers. A simple calculation shows that the controller gives the desired results. Let $p = d/dt$ be the differential operator. The following equations follow from the block diagram:

$$u = Ke + I$$

$$I = \frac{1}{1 + pT_i} u$$

The second equation implies that

$$T_i \frac{dI}{dt} + I = u = Ke + I$$

where the second equality follows from Equation 2.5. Hence,

$$T_i \frac{dI}{dt} = Ke$$

which shows that the controller in Figure 2.2 is, in fact, a PI controller.

PID Control

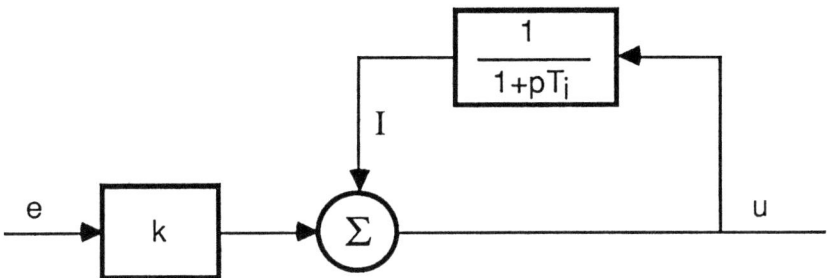

Figure 2.2
Interpretation of Integral Action as Automatic Reset

Derivative Action

The purpose of the derivative action is to improve the closed-loop stability. The instability mechanism, roughly speaking, can be described as follows. Because of the process dynamics it will take some time before a change in the control variable is noticeable in the process output. The action of a controller with proportional and derivative action may be interpreted as if the control is made proportional to the *predicted* process output, where the prediction is made by extrapolating the error by the tangent to the error curve (see Figure 2.3).

The prediction time is T_d.

Modifications of the Derivative Term

In the textbook algorithm, the derivative term is given as

$$D = KT_d \frac{de}{dt} = KT_d \left[\frac{dr}{dt} - \frac{dy}{dt} \right]$$

Set point r is normally constant with abrupt changes. It will thus normally not contribute to the derivative term. Moreover, the term dr/dt will change drastically when the set point is changed. For this reason it is common practice to apply the derivative action only to the process output. The derivative term is therefore implemented as

$$D = -KT_d \frac{dy}{dt} \tag{2.6}$$

PID Control

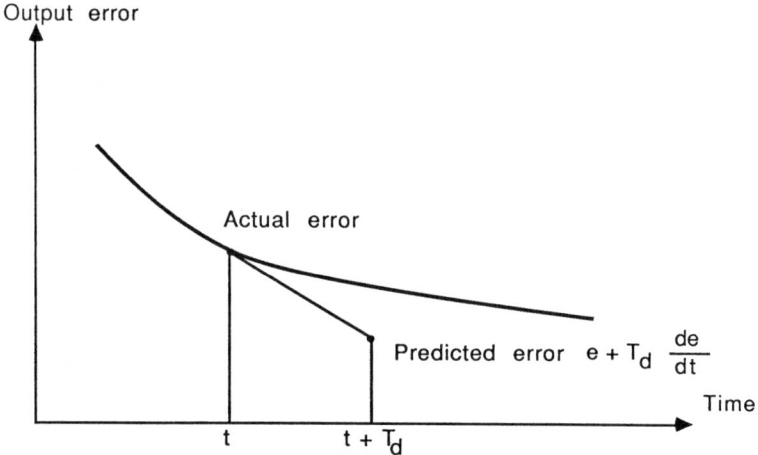

Figure 2.3
Interpretation of Derivative Action as Predictive Control

Limitation of the Derivative Gain

The derivative action may result in difficulties if there is high frequency measurement noise. A sinusoidal measurement noise

$$n = a \sin \omega t$$

will give the following contribution to the control signal:

$$u_n = KT_d \frac{dn}{dt} = aKT_d\omega \cos \omega t$$

The amplitude of the control signal can thus be arbitrarily large if the noise has a sufficiently high frequency (ω). The high frequency gain of the derivative term is therefore limited to avoid this difficulty. This can be done by implementing the derivative term as

$$\frac{T_d}{N} \frac{dD}{dt} + D = - KT_d \frac{dy}{dt} \tag{2.7}$$

It follows from this equation that the modified derivative term can be represented by the following operator:

$$D = - \frac{pKT_d}{1 + pT_d/N} y$$

PID Control

The modification can thus be interpreted as the ideal derivative filtered by a first-order system with the time constant T_d/N. The approximation will act as a derivative for low frequency signal components. The gain is, however, limited to N. This means that high frequency measurement noise is amplified most by a factor N.

Alternative Representations

The textbook PID algorithm (Equation 2.1) is sometimes replaced by the following controller:

$$G(p) = K' \left(1 + \frac{1}{pT_i'}\right)(1 + pT_d') \qquad (2.8)$$

This controller can always be represented in the form of Equation 2.1, where the coefficients are given by

$$K = K' \frac{T_i' + T_d'}{T_i'}$$

$$T_i = T_i' + T_d'$$

$$T_d = \frac{T_i' T_d'}{T_i' + T_d'}$$

A controller of the form (Equation 2.8) that corresponds to Equation 2.1 can be found only if

$$T_i > 4T_d$$

Then,

$$K' = \frac{K}{2}\left[1 + \sqrt{1 - 4T_d/T_i}\right]$$

$$T_i' = \frac{T_i}{2}\left[1 + \sqrt{1 - 4T_d/T_i}\right]$$

$$T_d' = \frac{T_i}{2}\left[1 + \sqrt{1 - 4T_d/T_i}\right]$$

Equation 2.1 is thus more general and we will stick to that in the future. It is, however, claimed that the form of Equation 2.8 is sometimes easier to tune manually.

PID Control

2.3 INTEGRAL WINDUP

Although many aspects of a control system can be understood based on linear theory, some nonlinear effects must be accounted for. All actuators have limitations: a motor has limited speed, a valve cannot be more than fully open or fully closed, etc. When a control system operates over a wide range of operating conditions, it may happen that the control variable reaches the actuator limits. When this happens the feedback loop is effectively broken because the actuator will remain at its limit independently of the process output. If a regulator with integrating action is used, the error will continue to be integrated. This means that the integral term may become very large or, colloquially, it "winds up". It is then required that the error change sign for a long period before things return to normal. The consequence is that any controller with integral action may give large transients when the actuator saturates.

The windup phenomenon is illustrated in Figure 2.4, which shows control of a process with a PI controller. The initial set point change is so large that the actuator saturates at the high limit. The integrator increases initially

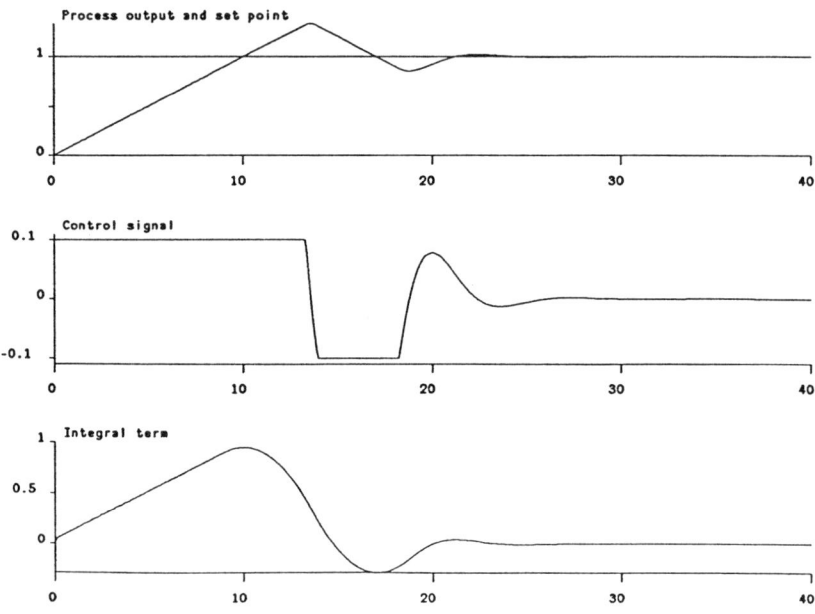

Figure 2.4
Illustration of Integral Windup

PID Control

because the error is positive, and it reaches its largest value at time $t = 10$ when the error goes through zero. The output remains saturated at this point because of the large value of the integral, and it does not leave the saturation limit until the error has been negative for a sufficiently long time to let the integral part come down to a small level. The net effect is a large overshoot, which is clearly noticeable in the figure. Integral windup may occur in connection with large set point changes or it may be caused by large disturbances or equipment malfunctions. Windup can also occur when selectors are used so that two controllers serve one actuator.

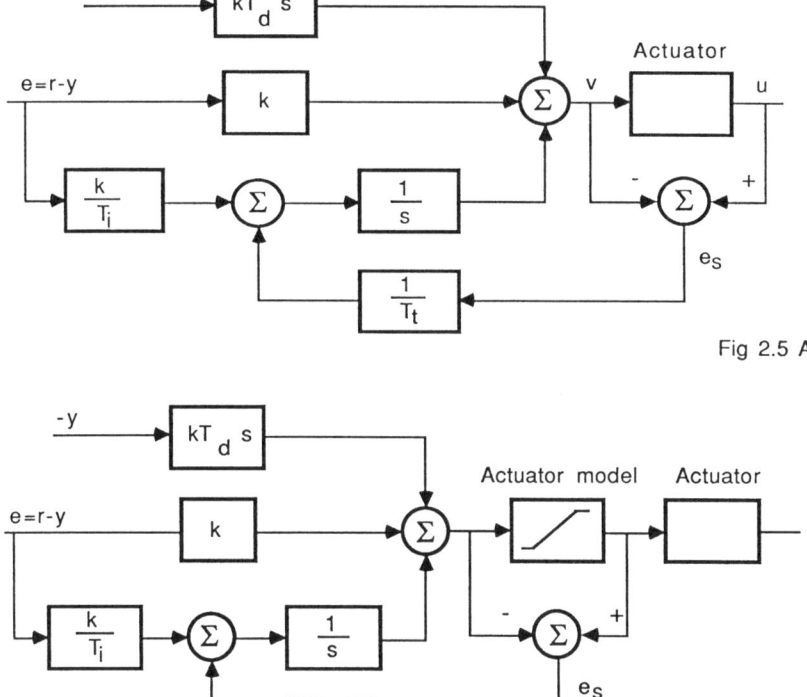

Fig 2.5 A

Fig 2.5 B

Figure 2.5
Controller with Anti-Windup
(A system where the actuator output is measured is shown in A and a system where the actuator output is estimated from a mathematical model is shown in B.)

PID Control

There are several ways to avoid integral windup. A convenient way is shown in Figure 2.5. An extra feedback path is provided in the controller by measuring the actual actuator output and forming an error signal (e_s) as the difference between the output of the controller (v) and the actuator output (u). Signal e_s is fed to the input of the integrator through gain $1/T_t$. The signal is zero when there is no saturation. It will thus not have any effect on the normal operation when the actuator does not saturate. When the actuator saturates, the feedback signal will, however, attempt to drive the error (e_s) to zero. This means that it attempts to drive the integrator to a value such that the controller output is exactly at the saturation limit. This will clearly prevent the integrator from winding up. The rate at which the controller output is reset is governed by the feedback gain, $1/T_t$, where T_t can be interpreted as the time constant, which determines how quickly the integral is reset. We call this the tracking time constant.

It frequently happens that the actuator output can not be measured. The anti-windup scheme just described can be applied by incorporating a mathematical model of the saturating actuator, as is illustrated in Figure 2.5B.

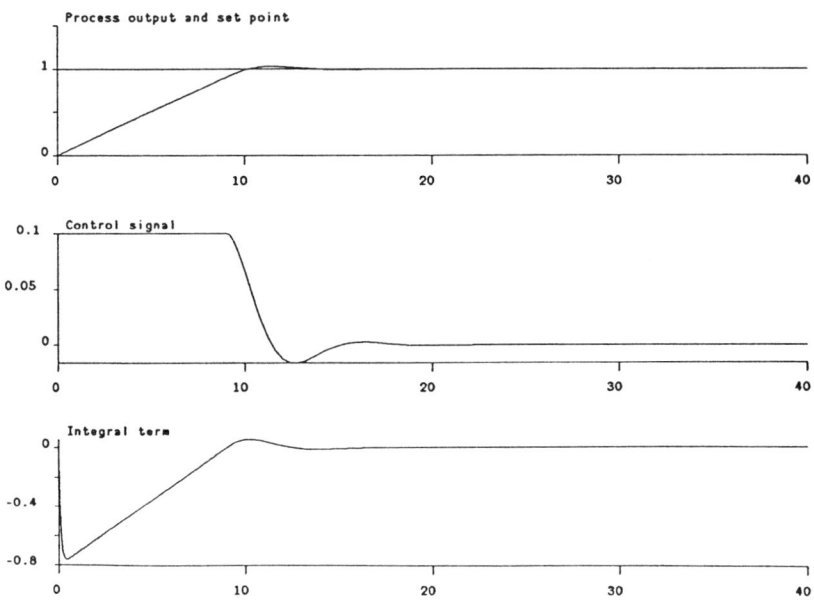

Figure 2.6
Controller with Anti-Windup Applied to the System of Figure 2.4

PID Control

Figure 2.6 shows what happens when a controller with anti-windup is applied to the system simulated in Figure 2.4. Notice that the output of the integrator is quickly reset to a value such that the controller output is at the saturation limit, and the integral has a negative value during the initial phase when the actuator is saturated. This behavior is drastically different from that in Figure 2.4, where the integral has a positive value during the initial transient. Also notice the drastic improvement in performance compared to the ordinary PI controller used in Figure 2.4.

The effect of changing the values of the tracking time constant is illustrated in Figure 2.7. It may thus seem advantageous to always choose a very small value of the time constant because the integrator is then reset quickly. However, some care must be exercised when introducing anti-windup in

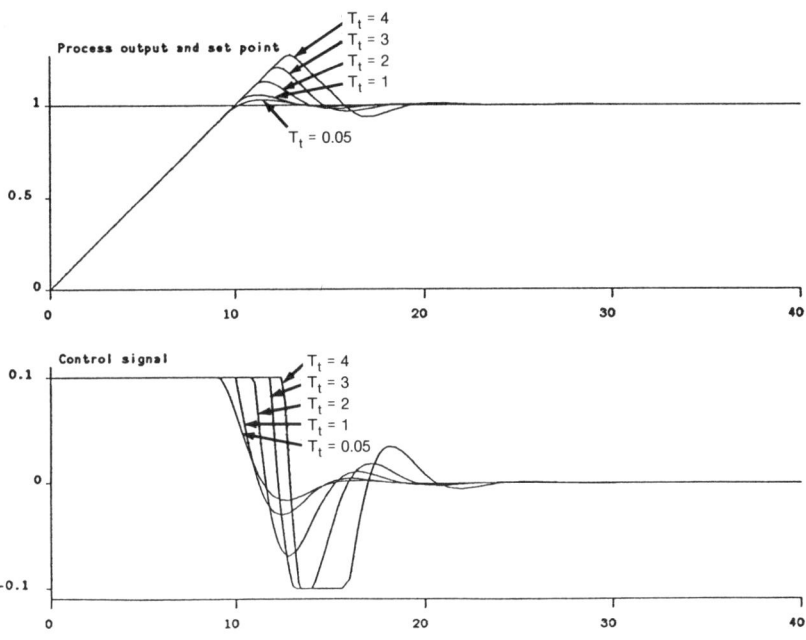

Figure 2.7
The Step Response of the System in Figure 2.6 for Different Values of the Tracking Time Constant T_t

PID Control

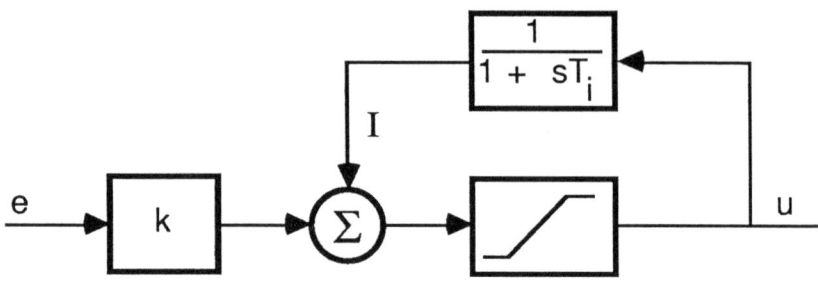

Actuator or actuator model

Figure 2.8
How to Provide Anti-Windup in the Controller in Figure 2.2
Where Integral Action is Generated as Automatic Reset

systems with derivative action. If the time constant is chosen too small, spurious errors can cause saturation of the output, which accidentally resets the integrator.

A similar device for avoiding windup can be applied to the controller shown in Figure 2.2 by incorporating a model of the saturation as shown in Figure 2.8. Notice that in this implementation the tracking time constant T_t is the same as the integration time T_i.

2.4 REFERENCE VALUES

A very common form of control system is shown in Figure 2.9. The system is characterized by forming an error that is the difference between the set point and the process output. The controller generates a control signal by operating on the error. This control signal is then applied to the plant. Such a system is called a "system with error feedback" because the controller operates on the error signal. The discussion of derivative action pointed out that it was advantageous not to let the derivative action act on the reference value. This means that the control law is not based on pure error feedback. A straightforward extension of this idea is to choose a PID controller of the form

PID Control

$$u = K\left[e_p + \frac{1}{T_i}\int_0^t e(s)\,ds + T_d \frac{de_d}{dt}\right] \quad (2.9)$$

where the error in the proportional part is

$e_p = br - y$

and the error in the derivative part is

$e_d = -y$

as before. The error in the integral part must be given as the true control error

$e = r - y$

to avoid steady-state errors. The controllers obtained for different values of b will respond to load disturbances and measurement noise in the same way. The response to set point changes will, however, depend on the value of b. This is illustrated in Figure 2.10, which shows the step response of a system with PI control of an integrator for different values of b. Notice that in the controller obtained for $b = 0$ the reference value is introduced only in the integral part.

In general, a control system has many different requirements. It should have good transient response to set point changes, and it should reject load disturbances and measurement noise. For a system with error feedback only, an attempt is made to satisfy all demands with the same mechanism.

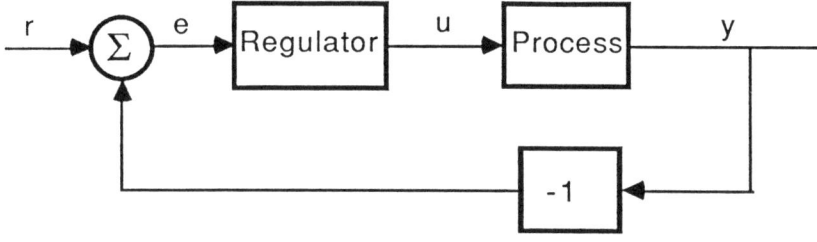

Figure 2.9
A Simple Feedback System with Error Feedback

PID Control

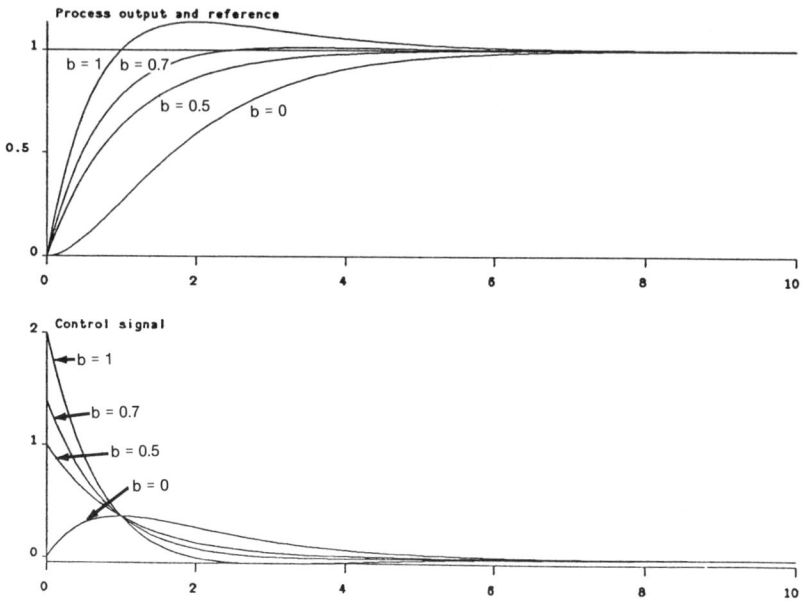

Figure 2.10
Effect of the Parameter b on the Step Response of a Closed-Loop System

(Such systems are also called one degree of freedom systems). By having different signal paths for the set point and the process output (two degrees of freedom systems), there is more flexibility to satisfy the design compromise. This is carried much further in more sophisticated control systems.

2.5. DIGITAL IMPLEMENTATION

PID controllers were originally implemented using analog techniques. Early systems used pneumatic relays, bellows, and needle valve constrictions. Electric motors with relays and feedback circuits and operational amplifiers were used later. Many of the features like anti-windup and derivation of process output were incorporated as "tricks" in the implementation. It is now common practice to implement PID controllers using microprocessors, and some of the old tricks have been rediscovered. Several issues must be considered in connection with digital implementations. The most important ones have to do with sampling, discretization, and quantization.

PID Control

Sampling

When a digital computer is used to implement a control law, all signal processing is done at discrete instances of time. The sequence of operations is as follows:

(1) Wait for clock interrupt
(2) Read analog input
(3) Compute control signal
(4) Set analog output
(5) Update controller variables
(6) Go to 1

The control actions are thus based on the values of the process output at discrete times only. This procedure is called *sampling*. The normal case is are sampled periodically with period h. The sampling mechanism introduces some unexpected phenomena, which must be taken into account in a good digital implementation of a PID controller. To explain these, consider the signals

$$s(t) = \cos(n\omega_s t \pm \omega t)$$

and

$$s_a(t) = \cos(\omega t)$$

where $\omega_s = 2\pi/h$ [rad/s] is the sampling frequency. Well-known formulas for the cosine function imply that the values of the signals at the sampling instants [kh, $k = 0, 1, 2,\ldots$] have the property

$$s(kh) = \cos(nkh\omega_s \pm \omega kh) = \cos(\omega kh) = s_a(\omega kh)$$

The signals s and s_a thus have the same values at the sampling instants. This means that there is no way to separate the signals if only their values at the sampling instants are known. Signal s_a is therefore called an *alias* of signal s. This is illustrated in Figure 2.11. A consequence of the aliasing effect is that a high frequency disturbance after sampling may appear as a low frequency signal. Assume, for example, that the sampling period is 18 ms. A 50-Hz sinusoidal disturbance will then, after sampling, appear as a sinusoid with the frequency

$$f_a = 50 - \frac{1}{0.018} = 5.6 \text{ Hz}$$

PID Control

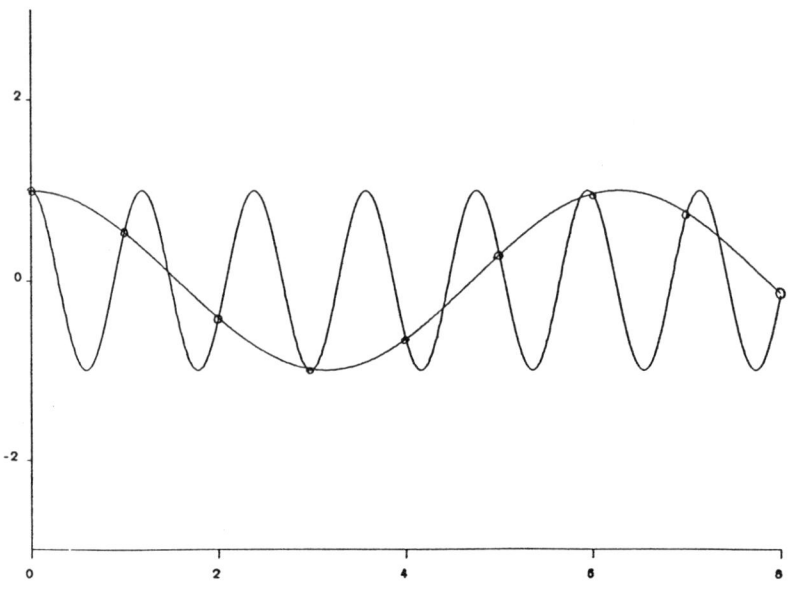

Figure 2.11
Illustration of the Aliasing Effect

Figure 2.12 shows all the aliases of a given frequency.

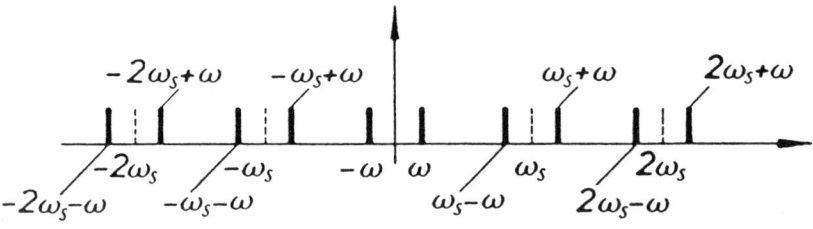

Figure 2.12
Aliases of the Frequency ω for a Sampling Frequency ω_s

Prefiltering

The alias effect can create significant difficulties if proper precautions are not taken. High frequencies, which in analog controllers normally are effectively eliminated by the low pass filtering, may because of aliasing appear as low frequency signals in the bandwidth of the sampled control system. To avoid these difficulties, an analog prefilter (which effectively eliminates all signal components with frequencies above half the sampling frequency) should be introduced. Such a filter is called an antialiasing filter. A second-order Butterworth filter is a common antialiasing filter. Higher-order filters are also used in critical applications. An implementation of such a filter using operational amplifiers is shown in Figure 2.13. The selection of the filter bandwidth is illustrated by the following example.

Example 2.1—Selection of Prefilter Bandwidth in a Second-Order Butterworth Filter

Assume it is desired that the prefilter attenuate signals by a factor of 16 at half the sampling frequency. If the filter bandwidth is ω_b and the sampling frequency is ω_s, we get

$$(\omega_s/2\omega_b)^2 = 16$$

Hence,

$$\omega_b = \frac{1}{8} \omega_s$$

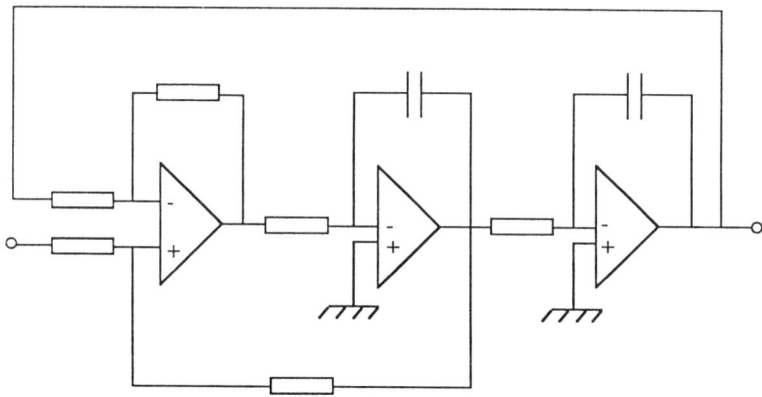

Figure 2.13
Circuit Diagram of a Second-Order Butterworth Filter

PID Control

Discretization

To implement a continuous time control law such as a PID controller on a digital computer, it is necessary to approximate the derivatives and the integral that appear in the control law. A few different ways to do this are presented below.

Proportional Action

The proportional term is

$$P = K(br - y)$$

This term is implemented simply by replacing the continuous variables with their sampled versions. Hence,

$$P(t_k) = K[br(t_k) - y(t_k)] \quad (2.10)$$

where $\{t_k\}$ denotes the sampling instants, i. e., the times when the computer reads the analog input.

Integral Action

The integral term is given by

$$I(t) = \frac{K}{T_i} \int_0^t e(s)\,ds$$

It thus follows that

$$\frac{dI}{dt} = \frac{K}{T_i} e$$

Approximating the derivative by a difference,

$$\frac{I(t_{k+1}) - I(t_k)}{h} = \frac{K}{T_i} e(t_k)$$

This leads to the following recursive equation for the integral term

$$I(t_{k+1}) = I(t_k) + \frac{Kh}{T_i} e(t_k) \quad (2.11)$$

Derivative Action

The derivative term is given by Equation 2.7, i. e.,

PID Control

$$\frac{T_d}{N}\frac{dD}{dt} + D = -KT_d\frac{dy}{dt} \tag{2.12}$$

There are several ways of approximating the derivative.

Forward Differences. Approximating the derivative by a forward difference gives

$$\frac{T_d}{N}\frac{D(t_{k+1}) - D(t_k)}{h} + D(t_k) = -KT_d\frac{y(t_{k+1}) - y(t_k)}{h}$$

This can be rewritten as

$$D(t_{k+1}) = \left[1 - \frac{hN}{T_d}\right]D(t_k) - KN[y(t_{k+1}) - y(t_k)] \tag{2.13}$$

Backward Differences. If the derivative in Equation 2.12 is instead approximated by a backward difference, the following is obtained:

$$\frac{T_d}{N}\frac{D(t_k) - D(t_{k-1})}{h} + D(t_k) = -KT_d\frac{y(t_k) - y(t_{k-1})}{h}$$

This can be rewritten as

$$D(t_k) = \frac{T_d}{T_d + Nh}D(t_{k-1}) - \frac{KT_dN}{T_d + Nh}[y(t_k) - y(t_{k-1})] \tag{2.14}$$

Tustin's Approximation. Yet another approximation proposed by Tustin is commonly used. This approximation is

$$D(t_k) = \frac{2T_d - hN}{2T_d + hN}D(t_{k-1}) - \frac{2KNT_d}{2T_d + hN}[y(t_k) - y(t_{k-1})] \tag{2.15}$$

Notice that all approximations have the same form, i. e.,

$$D(t_k) = a_i D(t_{k-1}) + b_i[y(t_k) - y(t_{k-1})] \tag{2.16}$$

but with different values of parameters a_i and b_i. The approximation of Equation 2.13 requires that $T_d > Nh/2$. The approximation becomes unstable for very small T_d. The other approximations are stable for all values of T_d. Notice, however, that Tustin's approximation and the forward difference approximation give negative values of a_i if $T_d < Nh/2$. This is undesirable because the approximation will then exhibit ringing. Hence only the approximation in Equation 2.14 will give good results for all values of T_d.

Figure 2.14 shows the phase curves for the approximations and for the corresponding continuous transfer function. Tustin's approximation gives

PID Control

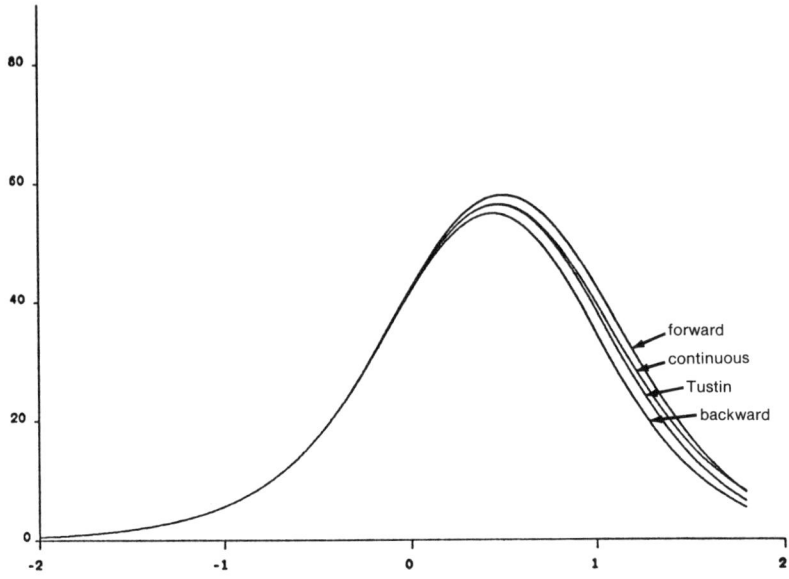

Figure 2.14
Phase Curves for Systems Obtained by Different Discretizations of the Derivative Term $pT_d/(1 = pT_d/N)$ with $T_d = 1$, $N = 10$ and a Sampling Period 0.02

the best agreement with the continuous time case, the backward approximation gives less phase advance, and the forward approximation gives more phase advance. The forward approximation is seldom used because of the problems with instability for small values of T_d. Tustin's algorithm is used quite frequently because of its close agreement with the continuous time case. The backward difference is used when an algorithm that is well behaved for small T_d is needed.

Incremental Form

The algorithms described so far are called positional algorithms because they give the output of the regulator directly. In digital implementations an incremental form of the algorithms is also used. This form is obtained by computing the time differences of the regulator output and adding the increments.

PID Control

$$\Delta u(t_k) = u(t_k) - u(t_{k-1}) = \Delta P(t_k) + \Delta I(t_k) + \Delta D(t_k)$$

The increments of the proportional part and the integral part are easily calculated from Equations 2.10 and 2.11.

$$\Delta P(t_k) = P(t_k) - P(t_{k-1}) = K[br(t_k) - y(t_k) - br(t_{k-1}) + y(t_{k-1})]$$

$$\Delta I(t_k) = I(t_k) - I(t_{k-1}) = \frac{Kh}{T_i} e(t_{k-1})$$

It was shown above that the derivative part can be calculated in various ways using different difference-approximation methods. Therefore, we use the unified description of Equation 2.16 to calculate the incremental derivative part.

$$\Delta D(t_k) = D(t_k) - D(t_{k-1}) = \frac{b_i}{1-a_i} [y(t_k) - 2y(t_{k-1}) + y(t_{k-2})]$$

One advantage with the incremental algorithm is that most of the computations are done using increments only. Short word-length calculations can often be used. It is only in the final stage where the increments are added that precision is needed. Another advantage with the incremental algorithm is that the regulator output is driven directly from an integrator. This makes it very easy to deal with windup and manual control. A problem with the incremental algorithm is that it cannot be used for controllers with P or PD action only because it will not keep the correct steady state. It is then necessary to use a modified algorithm when integral action is not used.

Quantization and Word Length

A digital computer allows only finite precision in the calculations. It is sometimes difficult to implement the integral term on computers with a short word length. The rounding off effect gives an offset, called integration offset. The difficulty is easily understood from Equation 2.11 for the integral term. The correction term $Kh/T_i * e$ is normally small in comparison to $I(t_k)$, and it may be rounded off unless the word length is sufficiently large. This rounding off effect gives an offset, called integration offset. To get a feel for the orders of magnitude involved, assume that all signals are normalized to have a largest magnitude of one. The correction term $Kh/T_i * e$ in Equation 2.11 then has the largest magnitude Kh/T_i. Let the sampling period h be 0.02 s, the integration time 20 min = 1200 s and the gain $K = 0.1$. Then,

$$\frac{Kh}{T_i} = 1.7 \; 10^{-6} = 2^{-19.2}$$

PID Control

To avoid rounding off the correction term, it is thus necessary to have a precision of at least 20 bits. More bits are required to obtain meaningful numerical values. The situation is particularly important when a stepping motor that outputs increments is used. It is then necessary to resort to special tricks to avoid rounding off the integral. One simple way is to use a longer sampling period for the integral term. For instance, if a sampling period of 1 s is used instead of 0.02 s in the previous example, a precision of 14 bits is sufficient.

Computer Code

As an illustration, the following is a computer code for a PID algorithm.

```
"Compute regulator coefficients
bi=K*h/Ti                         "integral gain
ad=(2*Td-N*h)/(2*Td+N*h)
bd=2*K*N*Td/(2*Td+N*h)            "derivative gain
aO=h/Tt

"Control algorithm
r=adin(ch1)                       "read set point from ch1
y=adin(ch2)                       "read process variable from ch2
P=K*(b*r-y)                       "compute proportional part
D=ad*D-bd*(y-yold)                "update derivative part
v=P+I+D                           "compute temporary output
u=sat(v,ulow,uhigh)               "simulate actuator saturation
daout(ch1)                        "set analog output ch1
I=I+bi*(r-y)+ao*(u-v)             "update integral
yold=y                            "update old process output
```

The computation of the coefficients must be done only when the controller parameters are changed. Precomputation of the coefficients ad, ao, bd, and bi saves computer time in the main loop. The main program must be called once every sampling period. The program has three states: yold, I, and D. One state variable can be eliminated at the cost of a less readable code. Notice that the code includes derivation of the process output only, proportional action on part of the error only ($b \neq = 1$), and anti-windup.

2.6 WHEN CAN PID CONTROL BE USED?

The requirements on a control system may include many factors such as response to command signals, insensitivity to measurement noise and process variations, and rejection of load disturbances. The design of a control system also involves aspects of process dynamics, actuator saturation, and disturbance characteristics. It may, therefore, seem surprising that a controller as simple as the PID regulator can work so well. The general empirical observation is that most industrial processes can be controlled reasonably well with PID control provided that the demands on the performance of the control are not too high. The following paragraphs delve further into this issue by first considering cases where PID control is sufficient and then discussing some generic problems where more sophisticated control is advisable.

When Is PI Control Sufficient?

Derivative action is frequently not used. It is an interesting observation that many industrial controllers only have PI action and that in others the derivative action can be (and frequently is) switched off in many control loops. It can be shown that PI control is adequate for all processes where the dynamics are essentially of the first order (level controls in single tanks, stirred tank reactors with perfect mixing, etc). It is fairly easy to find out if this is the case by measuring the step response or the frequency response of the process. If the step response looks like that of a first-order system or, more precisely, if the Nyquist curve lies in the first and the fourth quadrants only, then PI control is sufficient. Another reason is that the process has been designed so that its operation does not require tight control. Then even if the process has higher order dynamics, what it needs is an integral action to provide zero steady-state offset and an adequate transient response by proportional action.

When Is PID Control Sufficient?

Similarly, PID control is sufficient for processes where the dominant dynamics are of the second order, which is more difficult to establish. A frequency response measurement is one possibility. If the frequency response is monotone with a phase lag of less than 180 deg, then the system is of second order.

A typical case of derivative action improving the response is when the dynamics are characterized by time constants that differ in magnitude. Derivative action can then profitably be used to speed up the response. Temperature control is a typical case. Derivative control is also beneficial when tight control of a higher order system is required. The higher order dynamics would limit the amount of proportional gain for good control. With a derivative action, improved damping is provided, and hence a higher proportional gain can be used to speed up the transient response.

When Is More Sophisticated Control Needed?

Control of systems with a dominant time delay are notoriously difficult. It is also a topic on which there are many different opinions concerning the merit of PID control. There seems to be a general agreement that derivative action does not help much for processes with dominant time delays. For open-loop stable processes, the response to command signals can be improved substantially by introducing dead time compensation. The load disturbance rejection can also be improved to some degree because a dead time compensator allows a higher loop gain than a PID controller. Systems with dominant time delays are thus candidates for more sophisticated control.

Example 2.2—Dead time compensation.

Consider a process described by the equation

$$\frac{dy(t)}{dt} = -0.5y(t) + 0.5u(t-4)$$

The process has a time constant of 2 and a time delay of 4 time units. This process was first controlled by a PI controller with a gain of 0.2 and an integral time of 2.5 (see Figure 2.15). The figure also shows the properties of the control obtained with a Smith predictor. The response to set point changes is much improved, while the difference is less for the load disturbance. When dead time compensation was used, the gain in the PI controller was increased to $k = 1$, and the integral time was $T_i = 1$.

Systems with oscillatory modes that occur when there are inertias and compliances is another case where PID control is not sufficient. There are several approaches to systems of that type. In the so-called notch filter approach, no attempt is made to damp the oscillatory modes, but an effort is made to reduce the signal transmission through the regulator by a filter that drastically reduces signal transmission at the resonant frequency. A PID controller may be used when there is only one dominant oscillatory

PID Control

mode. Notch filter action can be achieved by a judicious choice of the controller parameters. In this case, parameters T_i and T_d should, however, be chosen so that the numerator has complex roots. The factored form in Equation 2.8 does not work in this case.

For some systems with large parameter variations it is possible to design linear controllers that allow operation over a wide parameter range. Such controllers are, however, often of high order.

The control of process variables that are closely related to important quality variables may be of a significant economic value. In such control loops it is frequently necessary to select the controller with respect to the disturbance characteristics. This often leads to strategies that are not of the PID type. These problems are often associated with time delays.

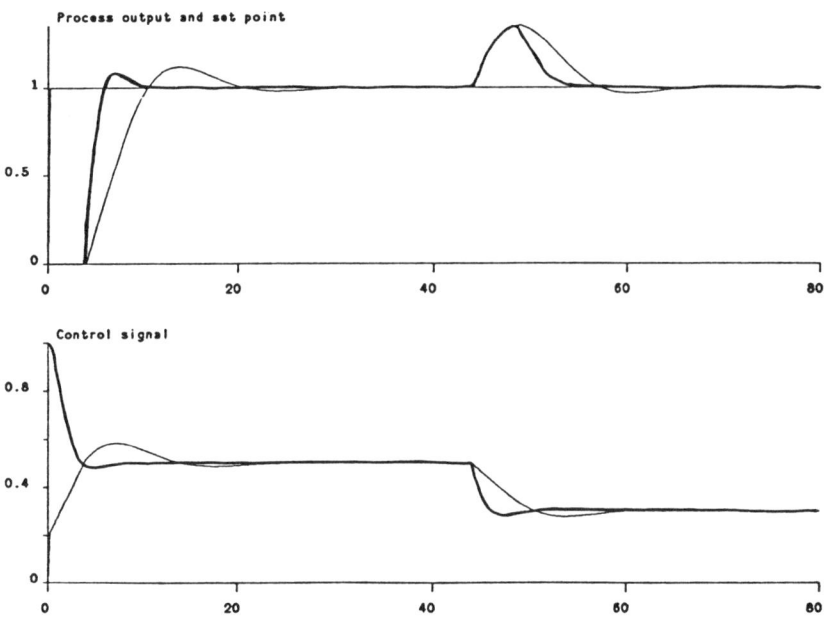

Figure 2.15
PI Control of the System in Example 2.2 with (thick line)
and without (thin line) Dead Time Compensation

PID Control

A general controller attempts to model the disturbances acting on the system. Since a PID controller has limited complexity, it cannot model complex disturbance behavior in general nor periodic disturbances in particular.

2.7 CONCLUSIONS

A detailed presentation of the PID algorithm has been given. Several modifications of the "textbook" version must be made to obtain a practical, useful controller. Problems that must be handled are, for example, integral windup and introduction of reference values. In a computer implementation, a discrete version of the PID algorithm is needed. Several methods to derive discrete PID algorithms have been described. Additional problems due to the sampling procedure must be handled, such as the design of a prefilter to avoid aliasing. A discussion of the limitations of the PID algorithm and a characterization of processes where the PID controller manages to perform the control have also been given.

Process Dynamics

3

3.1 INTRODUCTION

One method of tuning a controller is to first determine a model for the process dynamics and then calculate the controller parameters using some design method. Many tuning techniques are based on this approach. When PID controllers are tuned empirically, the process dynamics are also obtained indirectly. An understanding of the techniques for determining process dynamics is a good background for understanding controller tuning. This chapter summarizes such methods, which give models of varying complexity. The distinction is made between those methods where the transient is studied and the frequency response methods where the stationary response to sinusoidal signals is investigated. Transient response methods (useful in obtaining simple models of the process) are presented. The frequency response methods described can be used to obtain both simple models and more detailed descriptions. Methods based on estimation of parametric models are more complex than the transient and frequency response methods, but they are not limited to any specific characteristic of the input signal.

3.2 TRANSIENT RESPONSE

The dynamics of a process can be determined from the response of the process to pulses, steps, ramps, or other deterministic signals. If the system

Process Dynamics

is linear, at rest before the input is applied, and if there are no measurement errors, the process dynamics are, in principle, uniquely given from such a transient response experiment. In practice it is, however, difficult to ensure that the system is at rest. There will also be measurement errors, so the transient response method is, in practice, limited to the determination of simple models. Such models are, however, often sufficient for PID control. The methods are also very simple to use.

Step Response

Assuming a control loop with a regulator, the step response can easily be determined as follows. Wait until the process is at rest. Set the regulator to manual. Change the control variable rapidly, e.g., through the use of increase/ decrease buttons. Record the process variable and scale it by dividing by the change in the control variable.

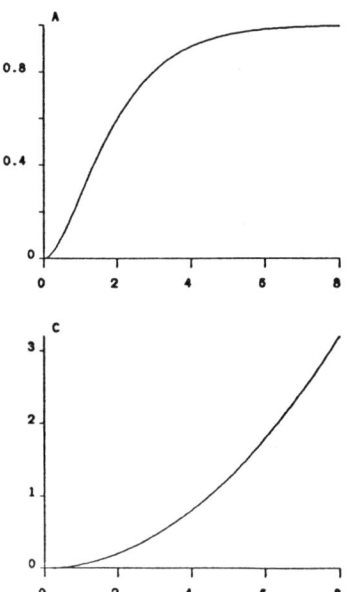
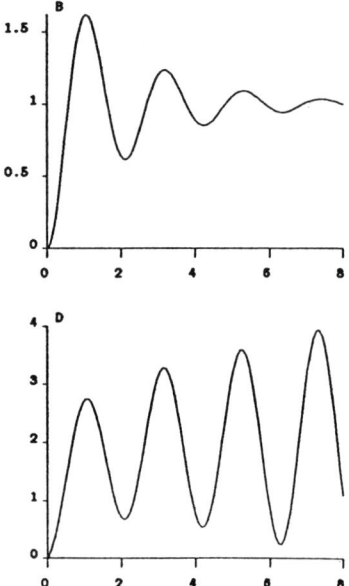

Figure 3.1
Open-Loop Step Responses

Process Dynamics

Examples of open-loop step responses are shown in Figure 3.1. Many properties of the system can be read directly from the step response. The systems in Figures 3.1A and B are stable, but the systems in Figures 3.1C and D are unstable. The systems in Figures 3.1B and D are oscillatory. A system can be shown to be linear by determining the step response for different input signal amplitudes: for a linear system the shape of the step response does not depend on the amplitude of the input signal. The step response is a convenient way to characterize process dynamics because of its simple physical interpretation. Many tuning methods are based on it. A formal mathematical model can also be obtained from the step response. General methods for the design of control systems can then be used. Some methods for determining parametric models from the step response are presented below.

Systems with Monotone Step Response

A step response is monotone if the step response does not decrease with time. The systems shown in Figures 3.1A and C have this property. Many industrial processes have monotone step responses. A simple model for such a system is given by the transfer function

$$G(s) = \frac{a}{s \cdot L} e^{-sL} \tag{3.1}$$

This corresponds to an integrator with delay. This model is characterized by the two parameters a and L, which are easily determined graphically from the step response (see Figure 3.2A). The model gives a reasonable description of the process behavior at the time scale of L, but it does not give a good description of the static (low frequency) behavior of the system. The model can, however, be used both for stable and unstable systems. A slightly more sophisticated model can be determined if the system is stable. This model has the transfer function

$$G(s) = \frac{k}{1 + sT} e^{-sL} \tag{3.2}$$

It is characterized by three parameters: the gain (k), the time constant (T), and the time delay (L). These parameters can be determined graphically by drawing the tangent to the step response that has the largest slope and finding the intersections of the tangent with the coordinate axes and the line $s(t) = k$ (see Figure 3.2B).

Process Dynamics

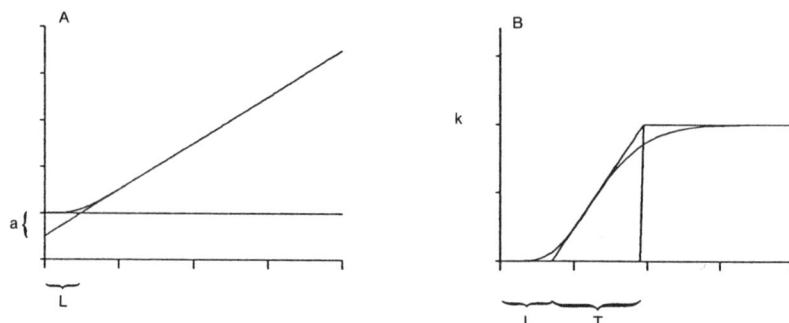

Figure 3.2
Graphical Determination of Mathematical Models for Systems with a Monotone Step Response

An even better approximation may be obtained by the transfer function

$$G(s) = \frac{k}{(1 + sT_1)(1 + sT_2)} e^{-sL} \tag{3.3}$$

This model has four parameters: the gain (k) the time constants (T_1 and T_2), and the time delay (L). Notice that the step response of a process that is distributed in nature or has multiple order dynamics increases very slowly initially. Such responses can often be approximated by Equations 3.2 or 3.3. The parameter L is then called the apparent time delay.

An Alternative Method

The construction in Figure 3.2 is sensitive because it relies on drawing a tangent to the step response. Another method, which is based on determination of areas, is shown in Figure 3.3. In this method the gain is first determined from the steady-state value of the step response. Area A_0 is then determined. The average residence time of the system is then

$$L + T = \frac{A_0}{k} \tag{3.4}$$

Process Dynamics

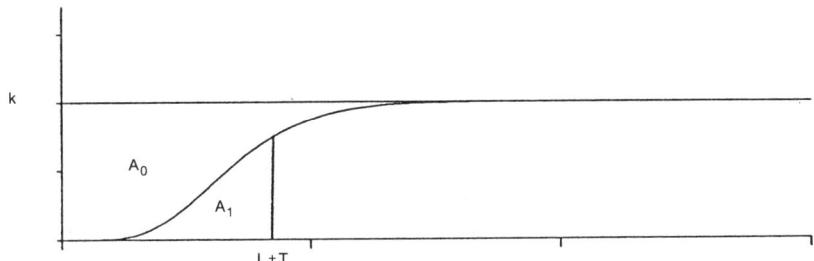

Figure 3.3
Graphical Determination of a Mathematical Model
for a Stable System with a Monotone Step Response

Area A_1, which is the area under the step response up to time $L + T$, is then determined and T is given by

$$T = \frac{A_1}{k} e^1 \tag{3.5}$$

This method is less sensitive to high frequency disturbances because it is based on area determination.

Oscillatory Systems

Oscillatory systems with step responses as shown in Figures 3.1B and D can be approximately modeled by the transfer function

$$G(s) = \frac{k\omega^2}{s^2 + 2\zeta\omega s + \omega^2} \tag{3.6}$$

This model has three parameters: the gain (k), the frequency (ω), and the relative damping (ζ). These parameters can be determined approximately from the step response as indicated in Figure 3.4. The period of the oscillation (T_p) and the damping (d) are first determined. Parameters ω and ζ are related to T_p and d as follows.

Process Dynamics

$$d = e^{-\frac{2\zeta\pi}{\sqrt{1-\zeta^2}}}$$

$$T_p = \frac{2\pi}{\omega\sqrt{1-\zeta^2}}$$

or

$$\begin{cases} \zeta = \left[\sqrt{1+(2\pi/\log d)^2}\right]^{-1} \\ \omega = \frac{2\pi}{T_p\sqrt{1-\zeta^2}} \end{cases} \quad (3.7)$$

A time delay can also be added. Notice also that the systems can be crudely approximated by Equation 3.1.

The Method of Moments

Equations 3.4 and 3.5 are special cases of a general method for determining the low frequency characteristics of a transfer function. To describe this method, let $h(t)$ be the impulse response and $G(s)$ the corresponding transfer function. The functions are related through

$$G(s) = \int_0^\infty e^{-st} h(t) dt$$

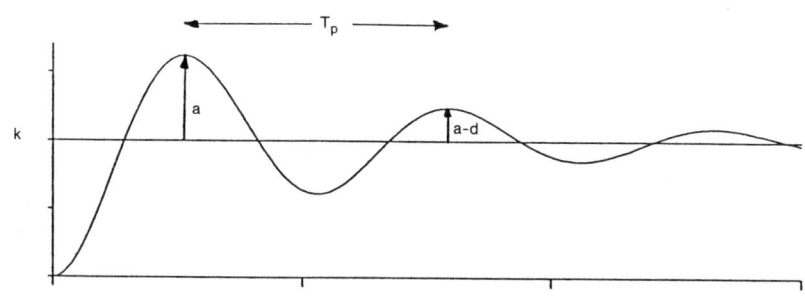

Figure 3.4
Graphical Determination of Mathematical Models for Systems with an Oscillatory Step Response

Taking derivatives with respect to s, the following are obtained:

$$G'(s) = -\int_0^\infty e^{-st} th(t)dt$$

$$G''(s) = \int_0^\infty e^{-st} t^2 h(t)dt$$

Hence,

$$G(0) = \int_0^\infty h(t)dt$$

$$G'(0) = -\int_0^\infty th(t)dt \qquad (3.8)$$

$$G''(0) = \int_0^\infty t^2 h(t)dt$$

The values of the transfer function and its derivatives for $\omega = 0$ can thus be determined from integrals of the impulse response.

A positive impulse response can be regarded as a mass distribution. The integrals above can then be interpreted as moments of the distribution, which explains the name of the method.

If the transfer function G is given by Equation 3.2, i.e.,

$$G(s) = \frac{k}{1 + sT} e^{-sL}$$

it follows that

$$G(0) = k$$
$$G'(0) = -k(T + L) \qquad (3.9)$$
$$G''(0) = k(2T^2 + 2TL + L^2)$$

For the transfer function (Equation 3.3), i.e.,

$$G(s) = \frac{k}{(1 + sT_1)(1 + sT_2)} e^{-sL}$$

it follows that

$$G(0) = k$$
$$G'(0) = -k(T_1 + T_2 + L) \qquad (3.10)$$
$$G''(0) = 2k(T_1^2 + T_2^2 + L^2 + T_1 T_2 + LT_1 + LT_2)$$

Process Dynamics

It is not necessary to measure impulse responses. Let U and Y be the Laplace transforms of an arbitrary input and the corresponding output, respectively. Then,

$Y(s) = G(s) U(s)$

$Y'(s) = G'(s) U(s) + G(s) U'(s)$

$Y''(s) = G''(s) U(s) + 2G'(s) U'(s) + G'(s) U''(s)$

Hence,

$Y(0) = G(0) U(0)$

$Y'(0) = G'(0) U(0) + G(0) U'(0)$ \hfill (3.11)

$Y''(0) = G''(0) U(0) + 2G'(0) U'(0) + G'(0) U''(0)$

The transfer function $G(0)$ and its derivatives $G'(0)$ and $G''(0)$ can thus be calculated from experiments with arbitrary inputs by calculating

$$U(0) = \int_0^\infty u(t)dt$$

$$U'(0) = -\int_0^\infty tu(t)dt$$

$$U''(0) = \int_0^\infty t^2 u(t)dt$$

and

$$Y(0) = \int_0^\infty y(t)dt$$

$$Y'(0) = -\int_0^\infty ty(t)dt$$

$$Y''(0) = \int_0^\infty t^2 y(t)dt$$

and using Equation 3.11.

Assuming that a unit step response is measured, the process gain $k = G(0)$ is then obtained, as the limit of the step response, as $t \to \infty$. Since the impulse response $h(t)$ is the derivative of a step response $s(t)$, it follows from Equation 3.8 that

Process Dynamics

$$G'(0) = -\int_0^\infty th(t)dt = -\int_0^\infty t\,\frac{ds(t)}{dt}\,dt = -\int_0^\infty t\,ds$$

The integral on the right-hand side can be interpreted as area A_0 shown in Figure 3.3. For a system characterized by a transfer function (Equation 3.2) we will then obtain Equation 3.4. The method of moments is, however, not restricted to step inputs. It can also be applied to arbitrary input signals, e.g., those obtained during closed-loop operation.

3.3 FREQUENCY RESPONSE

Consider a stable linear system. If the input signal to the system is a sinusoid, then the output signal will also be a sinusoid after a transient (see Figure 3.5). The output sinusoid will have the same frequency as the input signal; only the phase and the amplitude will be different. This means that under stationary conditions, the relationship between the input and the output can be described by two numbers: the quotient (a) between the input and the output amplitude and the phase shift (φ) between the input and the output signals. It is necessary to know the numbers a and φ for all frequencies (ω), that is, the functions $a(\omega)$ and $\varphi(\omega)$. It is convenient to view a and φ as the magnitude and the angle of a complex number

$$G(i\omega) = a(\omega)e^{i\varphi(\omega)} \tag{3.12}$$

The function $G(i\omega)$ is called the frequency response function of the system. The function $a(\omega) = |G(i\omega)|$ is called the amplitude function, and the function $\varphi(\omega) = arg(G(i\omega))$ is called the phase function.

The complex number $G(i\omega)$ can be represented by a vector with length $a(\omega)$ that forms angle $\varphi(\omega)$ with the x-axis, (see Figure 3.6). When the frequency goes from 0 to ∞, the endpoint of the vector describes a curve in the plane, which is called the frequency curve or the Nyquist curve.

The Nyquist curve gives a full description of the system. By sending sinusoids of different frequencies into the system, the Nyquist curve can thus be plotted, and a full description of the system can be obtained. This may be time consuming. Normally, it suffices to know only parts of the Nyquist curve. Of particular interest is the neighborhood of the lowest frequency where $G(i\omega)$ has a phase of $-180°$, called the crossover frequency (ω_c). The corresponding point on the Nyquist curve is called the critical point. The value of $G(i\omega_c)$ is all that is needed for some tuning methods.

Process Dynamics

Two methods for determining interesting points on the Nyquist curve are presented below. Both are based on the idea of using feedback to generate sinusoids having the appropriate frequency.

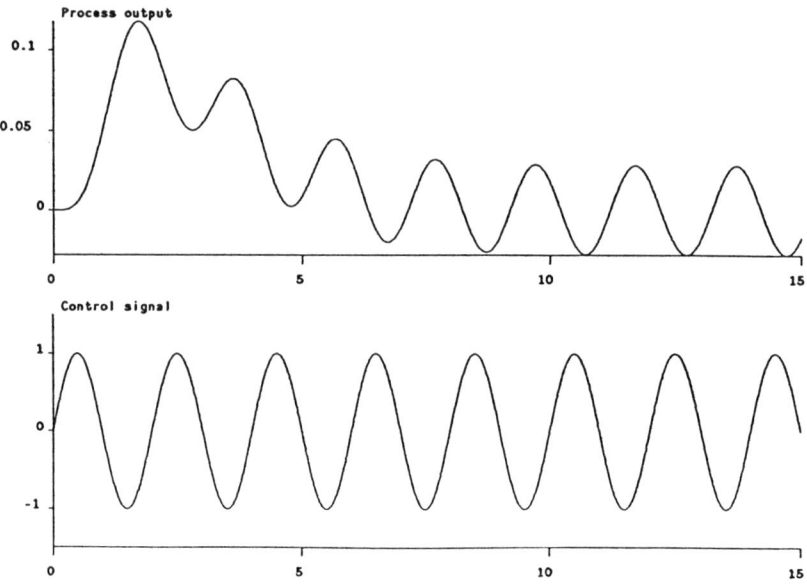

Figure 3.5
Response of a Linear Stable System to a Sinusoid

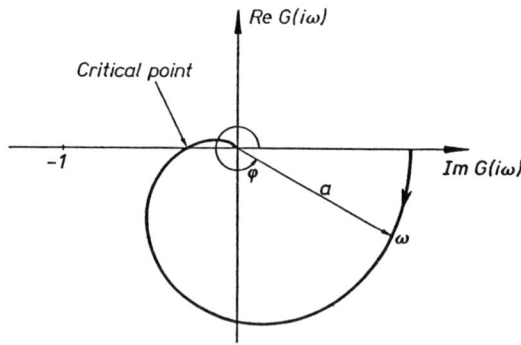

Figure 3.6
The Nyquist Curve of a System

Process Dynamics

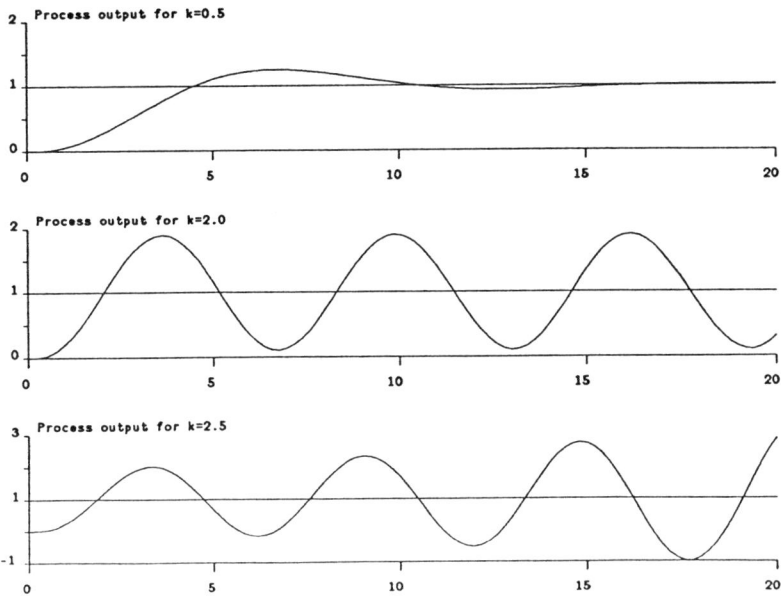

*Figure 3.7
Set Point and Process Variables
for a Closed-Loop System with Proportional Feedback*

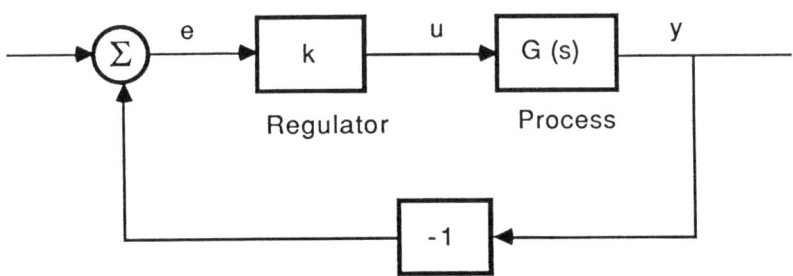

*Figure 3.8
Block Diagram of a Closed-Loop
System under Proportional Feedback*

The Ziegler-Nichols Frequency Response Method

Ziegler and Nichols have provided a method for determining the critical point on the Nyquist curve experimentally. The method is based on the observation that many systems can be made unstable under proportional feedback by choosing sufficiently high gain in the proportional feedback (see Figure 3.7). Assume that the gain is successfully adjusted so that the process is almost at the stability boundary. The control variable and the measured variable are then sinusoids with a phase shift of $-180°$ (see Figure 3.8). They are furthermore related by

$$u = -ky$$

because of the proportional feedback. Since the gain around the loop must be unity to maintain an oscillation,

$$k_c G(i\omega_c) = 1$$

where the gain, which brings the system to the stability limit, is called the ultimate gain (k_c). It follows from the above equation that

$$G(i\omega_c) = -\frac{1}{k_c} \qquad (3.13)$$

Several design methods based only on the knowledge of $G(i\omega_c)$ are given in Chapter 4.

The Ziegler-Nichols frequency response method has some advantages: it is based on a simple experiment and the process itself is used to find the ultimate frequency. It is, however, difficult to automate this experiment or perform it in such a way that the amplitude of the oscillation is kept under control. Operating the process near instability is also dangerous and needs management authorization in an industrial plant. It is, therefore, not useful for autotuning. Another method for automatic determination of specific points on the Nyquist curve is suggested below.

Relay Feedback

The method is based on the observation that the appropriate oscillation can be generated by relay feedback. The system is thus connected as shown in Figure 3.9. For many systems there will then be an oscillation (as shown in Figure 3.10) where the control variable is a square wave and the process

Process Dynamics

output is close to a sinusoid. Notice that the process input and output are out of phase and that the amplitude of the oscillation is proportional to the relay amplitude.

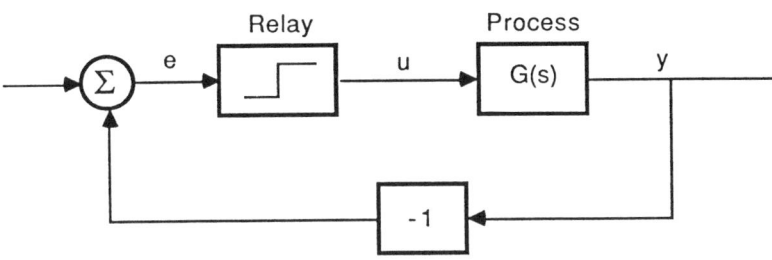

Figure 3.9
Block Diagram of a Process Under Relay Feedback

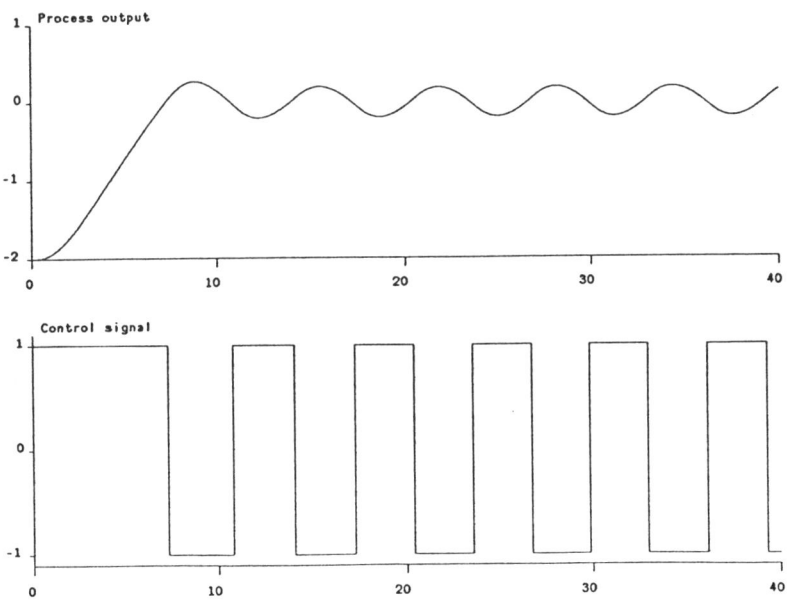

Figure 3.10
Relay Output u and Process Output y for a System Under Relay Feedback

Process Dynamics

To explain how the system works, assume that the relay output is expanded in a Fourier series and that the process attenuates higher harmonics effectively. If d is the relay amplitude, its first harmonic is $4d/\pi$. Let a be the amplitude of the oscillation in the process output. Then,

$$G\left(i\frac{2\pi}{t_c}\right) = -\frac{\pi a}{4d} \tag{3.14}$$

Notice that the relay experiment is easily automated. Since the amplitude of the oscillation is proportional to the relay output, it is easy to control it by adjusting the relay output.

Other Possibilities

The idea to determine process dynamics with relay feedback can be used in several different ways. A known linearity can be connected in series with the process. A relay followed by an integrator will, for example, admit determination of the point where the process has a phase shift of $-90°$. A relay with a differentiator admits determination of the point where the plant has a phase shift of $-270°$.

Describing Function Analysis

The describing function method is an approximate method that can be used to determine if there will be an oscillation or not if a nonlinear function is introduced in a control loop. To determine conditions for oscillation, the nonlinear block is described by a gain, $N(a)$, which depends on signal amplitude a at the input of the nonlinearity. This gain is called the describing function. If the process has the transfer function $G(i\omega)$, the condition for oscillation is simply given by

$$N(a)G(i\omega) = -1 \tag{3.15}$$

This equation is obtained by requiring that a sine wave with frequency ω should propagate around the feedback loop with the same amplitude and phase. Since N and G may be complex numbers, this gives two equations for determining a and ω. The equation can be solved graphically by plotting

Process Dynamics

$-1/N(a)$ in the Nyquist diagram. If the negative inverse of the describing function is drawn in the complex plane (as in Figure 3.11) together with the Nyquist curve of the linear system, an oscillation may occur if there is an intersection between the two curves. The amplitude and the frequency of the oscillation are the same as the parameters of the two curves at the intersection point. Therefore, measuring the amplitude and the period of the oscillation, the position of one point of the Nyquist curve can be determined.

The describing function, $N(a)$, for a relay is given by

$$N(a) = \frac{4d}{\pi a} \tag{3.16}$$

Since this function is real, the oscillation may occur where the Nyquist curve intersects the negative real axis. Thus, the conclusion is that the critical point, i.e., the intersection of the Nyquist curve with the negative real axis, is conveniently determined by a relay feedback experiment.

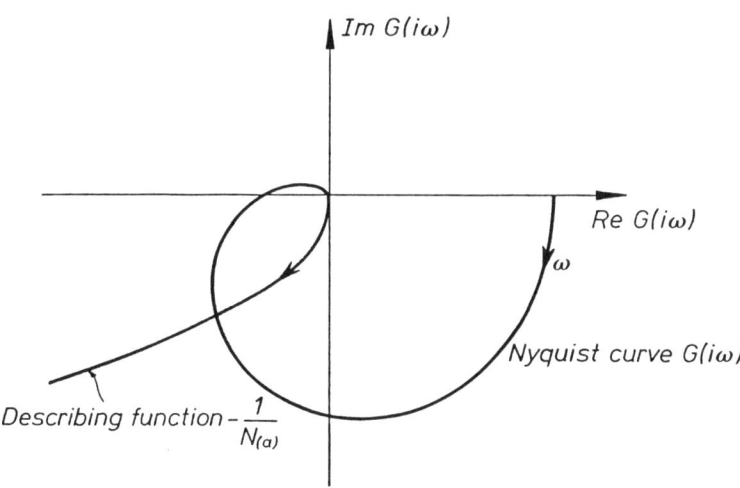

Figure 3.11
Determination of Possible Oscillations Using the Describing Function Method

Process Dynamics

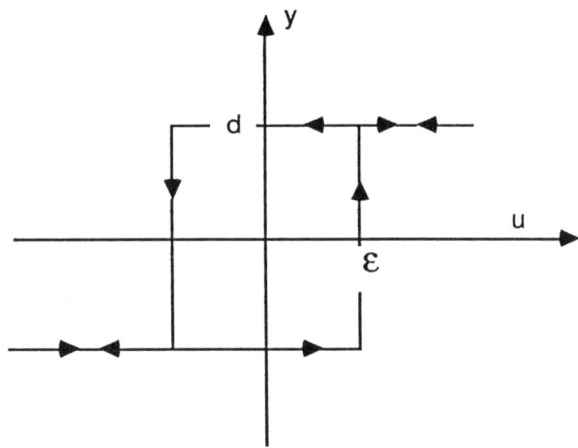

Figure 3.12
Input/Output Characteristics of a Relay with Hysteresis

A Relay with Hysteresis

There are advantages in having a relay with hysteresis instead of a pure relay. This can intuitively be understood as follows. With an ordinary relay, a small amount of noise can make the relay switch. By introducing hysteresis, the noise must be larger than the hysteresis width to make the relay switch (see Figure 3.12, which shows the input/output characteristics of a relay with hysteresis). The describing function approach will be used to investigate the oscillations obtained. The negative reciprocal of the describing function of such a relay is

$$-\frac{1}{N(a)} = -\frac{\pi}{4d}\sqrt{a^2 - \epsilon^2} - i\,\frac{\pi\epsilon}{4d} \tag{3.17}$$

where d is the relay amplitude and ϵ is the hysteresis width. This function can be regarded as a straight line parallel to the real axis, in the complex plane (see Figure 3.13).

By choosing the relation between ϵ and d, it is therefore possible to determine a point on the Nyquist curve with a specified imaginary part. Several points on the Nyquist curve are easily obtained by repeating the experiment with different relations between ϵ and d. This is useful for the

Process Dynamics

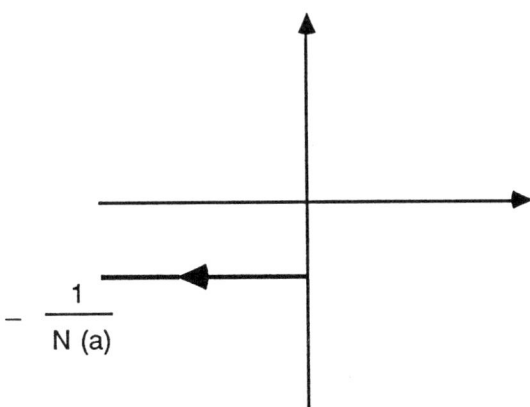

Figure 3.13
The Negative Reciprocal of the Describing
Function N(a) for a Relay with Hysteresis

dominant pole design described in Chapter 4. It is easy to control the amplitude of the limit cycle to a desired level by a proper choice of the relay amplitude.

Other Nonlinearities

It is possible to use other types of nonlinearities. If a nonlinearity with the describing function shown in Figure 3.14 is used, it is possible to determine a point on the Nyquist curve with a prescribed phase shift. Different nonlinearities can also be combined with linear links with known dynamics in series with the plant as discussed above.

3.4 PARAMETER ESTIMATION

A mathematical model of the process can also be obtained by fitting the parameters of a model to experimental data. For example, an attempt could be made to fit a model of the type in Equation 3.2 directly to observed input/output data. The advantage of such an approach is that any type of input/output data can be used. However, parameter estimation requires more computations than the methods discussed previously.

Process Dynamics

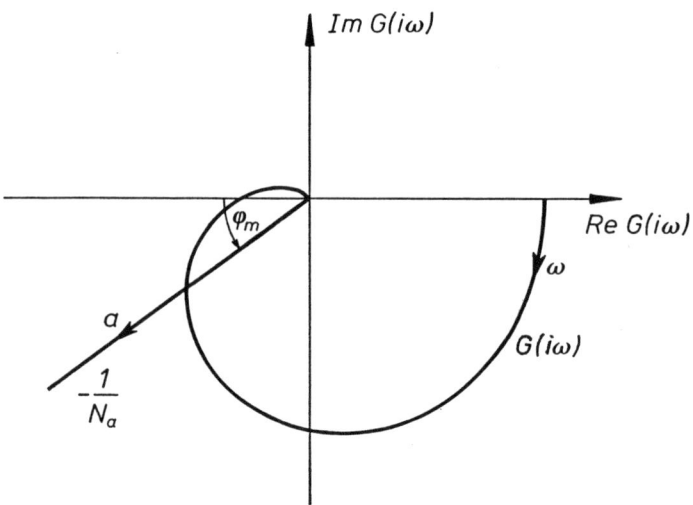

Figure 3.14
Nyquist Curve and Describing Function for a Nonlinearity That Admits Determination of a Point on $G(i\omega)$ with a Predetermined Phase Shift

Parametric Models

Since the calculations will typically be made using a digital computer, the input/output data will typically be sampled. It is then convenient to operate with a discrete time model based on signals that are sampled periodically. Moreover, if the experimental data is also computer-generated, it is reasonable to assume that the input to the plant is constant between the sampling instants. Let the sampling period be h. Assume that time delay L is less than h. The model (Equation 3.2) can then be described as

$$y(kh) = ay(kh - h) + b_1 u(kh - h) + b_2 u(kh - 2h) \qquad (3.18)$$

where

$$a = e^{-h/T}$$

$$b_1 = k\left[1 - e^{-\frac{h-L}{T}}\right]$$

$$b_2 = ke^{-h/t}[e^{L/t} - 1]$$

Process Dynamics

For arbitrary time delays L, the model becomes instead

$$y(kh) = ay(kh - h) + b_1 u(kh - nh) + b_2 u(kh - nh - h) \qquad (3.19)$$

where parameters a, b_1, and b_2 are given as above with $n = L$ div h and $\tau = L$ mod h replacing L. The model can be given a convenient representation by introducing a shift operator q, defined by

$$qy(kh) = y(kh+h)$$

The model (Equation 3.19) can then be written as

$$q^n (q - a) y(kh) = (b_1 q + b_2) u(kh)$$

If the complex variable z (similar to the Laplace transform variable s) is introduced, the process can also be described by the pulse transfer function:

$$H(z) = \frac{b_1 z + b_2}{z^n (z - a)} \qquad (3.20)$$

Notice that the transfer function is a ratio of two polynomials even if the corresponding physical process has time delays.

The discussion can be extended to systems of higher order, and the result is then an input/output relation of the form:

$$A(q)y(kh) = B(q)u(kh) \qquad (3.21)$$

where $A(q)$ and $B(q)$ are polynomials:

$$A(q) = q^n + a_1 q^{n-1} + \ldots + a_n$$
$$B(q) = b_1 q^{n-1} + b_2 q^{n-2} + \ldots + b_n$$

The corresponding transfer function is then

$$H(z) = \frac{B(z)}{A(z)}$$

Parameter Estimation

There are many ways to estimate the parameters of the discrete time model (Equation 3.21). A simple method is as follows. Assume that a sequence of input/output pairs ($\{u(kh), y(kh), k = 1, 2, \ldots, N\}$) have been

Process Dynamics

observed. The parameters can then be determined in such a way that Equation 3.21 fits the data as well as possible in the least squares sense. The square of the errors is

$$V(\theta) = \sum_{k=n+1}^{N} e^2(kh) \qquad (3.22)$$

where

$$e(kh + nh) = A(q)y(kh) - B(q)u(kh), \, k = 1, \ldots, N - n$$

Notice that the error is linear in parameters a_i and b_i of the model and that the sum of squares of the errors is a quadratic function. This means that the problem of minimizing the loss function can easily be made analytically. Rather than showing the solution to the optimization problem, a convenient way of computing the parameters recursively is presented below.

Recursive Computations

In a tuning experiment, the input/output data is normally obtained recursively. Since a new input/output pair is obtained each sampling period, it is convenient to compute the parameter estimates recursively. The formula for this follows, along with some shorthand for the purpose. All parameters are grouped together in the vector:

$$\theta = \begin{bmatrix} a_1 \\ a_2 \\ \vdots \\ a_n \\ \vdots \\ b_1 \\ b_2 \\ \vdots \\ b_n \end{bmatrix}$$

Also, the regression vector is introduced:

$$x_{k-1} = [-y(kh - h) \ldots -y(kh - nh) \, u(kh - h) \ldots u(kh - nh)]^T$$

Process Dynamics

The estimate can then be calculated recursively by

$$e_k = y(kh) - x_{k-1}^T \theta_{k-1} \tag{3.23A}$$

$$P_k = P_{k-1} - \frac{P_{k-1} x_{k-1} x_{k-1}^T P_{k-1}}{1 + x_{k-1}^T P_{k-1} x_{k-1}} \tag{3.23B}$$

$$\theta_k = \theta_{k-1} + P_k x_{k-1} e_k \tag{3.23C}$$

These equations have good physical interpretation. The new estimate (θ_k) is obtained by adding a correction term (Pxe) to the old estimate (θ_{k-1}). The correction term is a product of three quantities: P, x, and e. The error (e) is the difference between the last measurement ($y(kh)$) and the prediction ($x^T\theta$) of this measurement based on old estimates. Regression vector x can be interpreted as the gradient of the error with respect to the parameters.

Equation 3.23B may be interpreted as follows. Matrix P_k is proportional to the covariance matrix of the estimates; the last term in Equation 3.23B is the reduction in uncertainty due to the last measurement.

The equations have to be initialized. The initial value of parameter vector θ can be chosen as the best initial guesses of the parameters. The initial value of matrix P is typically chosen as the identity matrix multiplied by a large number.

Computer Code

Equations 3.23 for recursive estimation are an essential part of many schemes for automatic tuning. The following is a computer code that implements the algorithm.

{The recursive least squares algorithm}

```
e=y
for i=1 to 2*n do e=e-x[i]*θ[i]
```

{Compute estimator gain}

```
for i=1 to 2*n do
    begin
    s=0
    d=1
    for j=1 to 2*n do
        begin
        s=s+P[i,j]*x[j]
        d=d+s*x[j]
```

Process Dynamics

```
    end
    r[i]=s
end
```

{Update estimates}

for i=1 to 2*n do θ[i]=θ[i]+r[i]*e/d

{Update P matrix}

```
for i=1 to 2*n do
  begin
  for j=i to 2*n do P[i,j]=P[i,j]-r[i]*r[j]/d
  for j=i+1 to 2*n do P[j,i]=P[i,j]
  end
```

{Update x-vector}

for i=1 to 2*n-1 do x[2*n-i+1]=x[2*n-i]
x[1]=-y
x[n+1]=u

The code description is given in pidgin Pascal, and it is assumed that the variables have been properly declared. There are many refinements to the algorithm; for instance, its numerical properties can be improved by using a so-called square root algorithm. It is also common practice to bandpass filter the signals before introducing them into the algorithm to get rid of dc levels and high frequency disturbances. There are also many variations of the algorithm to discount past data. The code will, however, give an indication of the type of algorithms that are used in recursive parameter estimation.

3.5 CONCLUSIONS

A number of methods for determining the dynamics of a process have been presented in this chapter. Some are very simple: they are based on a direct measurement of the step or pulse response and simple graphical constructions. Others are based on the frequency response. It was shown that very useful information could be generated from relay experiments. Such experiments were particularly useful because the process is brought into self-oscillation at the ultimate frequency, which is of considerable interest for design of controllers. In later chapters, it will be demonstrated how the models can be used to design controllers.

Design of PID Controllers

4

4.1 INTRODUCTION

This chapter describes some methods for determining the parameters of a PID controller. The properties of the design methods will be illustrated using a fourth-order process model. The methods differ with respect to the knowledge of the process dynamics they require. A PI controller is described by two parameters (K and T_i) and a PID regulator by three or four parameters (K, T_i, T_d and N). In the classical Ziegler-Nichols methods, the dynamics are characterized by two parameters. In the step response method, they are taken from the step response. In the Ziegler-Nichols frequency response method, the parameters are the frequency where the open-loop dynamics have a phase shift of 180° and the gain at that frequency. An obvious extension of the frequency response method is to develop methods that are based on knowledge of the open-loop transfer function at two frequencies, i.e., four parameters. Another way to obtain a characterization of process dynamics with few parameters is, of course, to use low-order dynamic models with few parameters. Design methods based on dynamic models of first and second order are discussed. A corresponding treatment of discrete time models is also given. The discrete time models have the advantage that they can describe time delays using finite dimensional models. Many of the design methods described give good responses to load disturbances. The

Design of PID Controllers

response to command signals will, however, often show a significant overshoot. The nature of this problem is discussed, and it is shown that the difficulty is due to a deficiency of the conventional PID structure. A simple way to alleviate this problem is suggested. The different design methods are compared, and some insight into the sensitivity problem and the differences between PI and PID control are also given.

4.2 ZIEGLER-NICHOLS METHODS

Two classical methods were presented by Ziegler and Nichols in 1942. These methods are still widely used, either in their original form or in some modification.

Ziegler-Nichols Step Response Method

The first design method presented by Ziegler and Nichols is based on a registration of the open-loop step response of the system, which is characterized by two parameters (see Figure 4.1). The point where the slope of the step response has its maximum is first determined, and the tangent at this point is drawn. The intersections between this tangent and the coordinate axes give the two parameters a and L. In Chapter 3, a model of the process to be controlled was derived from these parameters. Ziegler and Nichols have given PID parameters directly as functions of a and L. These are given in Table 4.1. An estimate of the period T_p of the dominant dynamics of the closed-loop system is also given in the table.

Example 4.1—The Ziegler-Nichols step response method will be applied to the process

$$G_p(s) = \frac{1}{(1 + s)(1 + 0.2s)(1 + 0.05s)(1 + 0.01s)} \quad (4.1)$$

This process model is used as a test example throughout the chapter. Measurements on the step response give the parameters $a = 0.11$ and $L = 0.16$. The controller parameters can now be determined from Table 4.1. It follows that a PI controller should have the parameters $K = 8.2$ and $T_i = 0.48$. The parameters of a PID controller are $K = 10.9$, $T_i = 0.32$, and $T_d = 0.08$. Figure 4.2 shows the response of the closed-loop systems to a step command and a load disturbance.

Design of PID Controllers

Table 4.1
Recommended PID Parameters According to
Ziegler-Nichols Step Response Method

Controller	K	T_i	T_d	T_p
P	1/a			4L
PI	0.9/a	3L		5.7L
PID	1.2/a	2L	L/2	3.4L

Notice that the response of the PI controller is poorly damped but that the response of the PID controller is better. The overshoot in the response to the command signal is, however, excessive even for the PID controller.

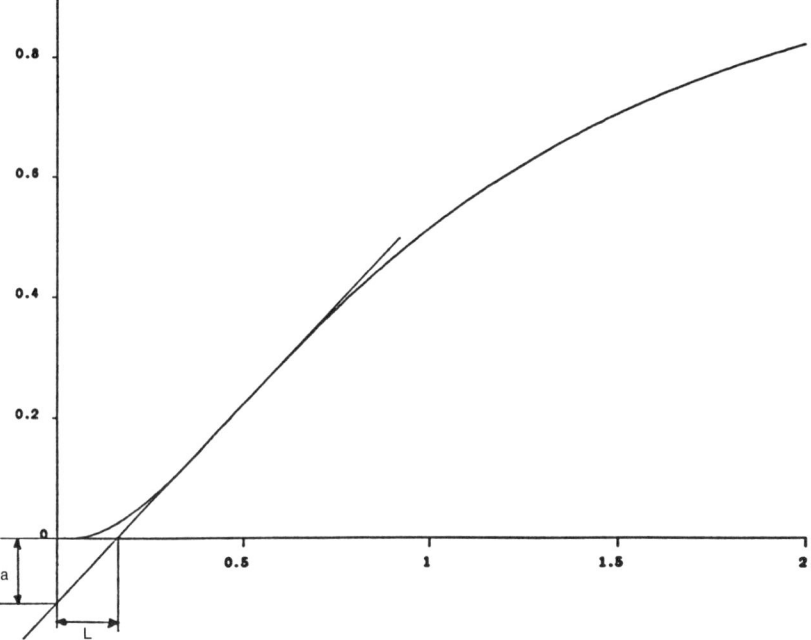

Figure 4.1
Characterization of a Step Response
Used in the Ziegler-Nichols Step Response Method

Design of PID Controllers

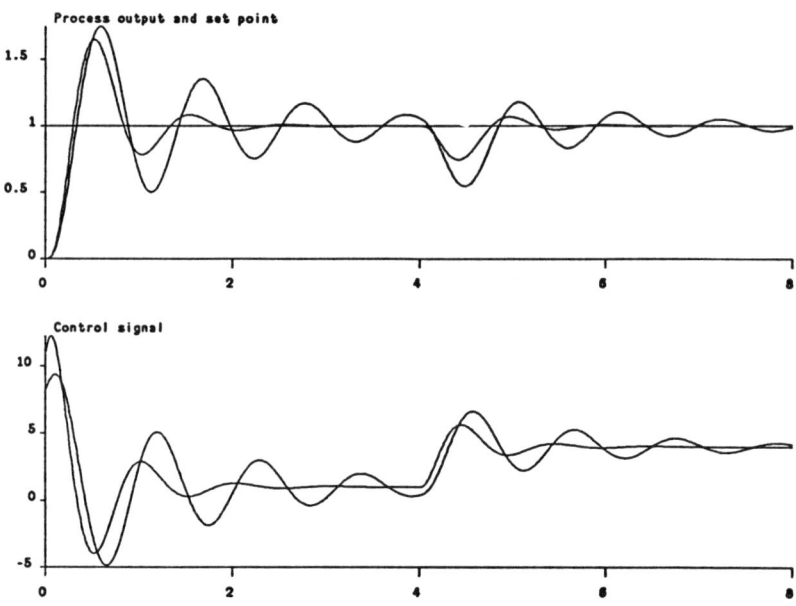

Figure 4.2
Step and Load Disturbance Response of the Process (Equation 4.1) Controlled by a PI Controller (thin lines) a PID Controller (thick lines) Tuned by the Ziegler-Nichols Step Response Method

Ziegler-Nichols Frequency Response Method

This method is also based on a very simple characterization of the process dynamics. The design is based on knowledge of the point on the Nyquist curve of the process transfer function G where the Nyquist curve intersects the negative real axis. For historical reasons this point is characterized by the parameters k_c and t_c, which are called the *ultimate gain* and the *ultimate period*. Section 3.2 described a method to obtain parameters k_c and t_c by increasing the gain in a proportional controller until the stability boundary is reached. The parameters can also be obtained using the relay feedback experiment presented in the same section. The Ziegler-Nichols design method gives simple formulas for the parameters of the controller in terms

Design of PID Controllers

of the ultimate gain and the ultimate period (see Table 4.2). An estimate of the period (T_p) of the dominant dynamics of the closed-loop system is also given in the table.

Table 4.2
Recommended PID Parameters According to
Ziegler-Nichols Frequency Response Method

Controller	K	T_i	T_d	T_p
P	0.5 k_c			t_c
PI	0.4 k_c	0.8 t_c		1.4 t_c
PID	0.6 k_c	0.5 t_c	0.12 t_c	0.85 t_c

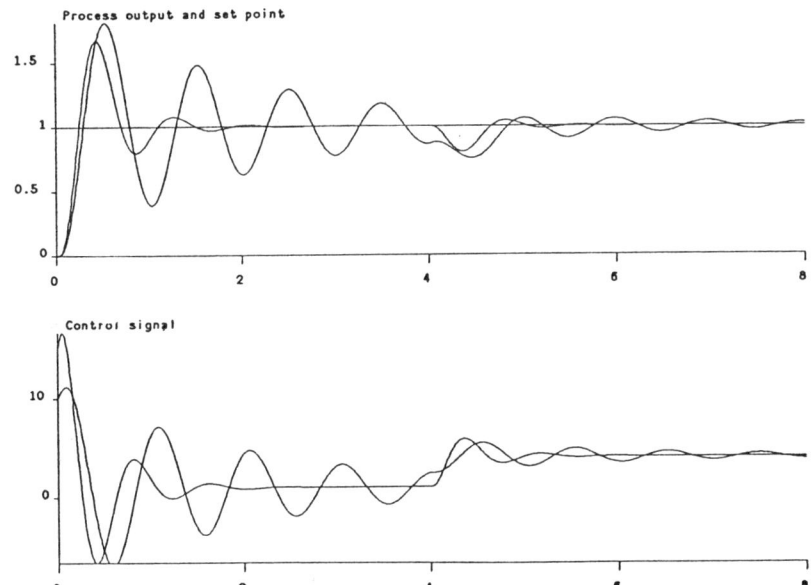

Figure 4.3
Step and Load Disturbance Response When the Process
(Equation 4.1) is Controlled by a PI Controller (thin lines) and a PID Controller (thick lines) Tuned by Ziegler-Nichols Frequency Response Method

Design of PID Controllers

Example 4.2—The process of Equation 4.1 has the ultimate gain $k_c \approx 25$ and the ultimate period $t_c \approx 0.63$. Table 4.2 gives the parameters $K = 10$ and $T_i = 0.50$ for a PI controller and $K = 15$, $T_i = 0.31$, and $T_d = 0.08$ for a PID controller. Figure 4.3 shows the closed-loop step and load disturbance responses when the controllers are applied to the Equation 4.1 process. The parameters and the performance of the controllers obtained with the frequency response method are quite close to those obtained by the step response method.

The Ziegler-Nichols tuning rules were originally designed to give systems with good responses to load disturbances. They were obtained by extensive simulations of many different systems. The design criterion was quarter amplitude damping. In Section 3.2, the relation between the damping (d) and the relative damping (ζ) is given as:

$$\zeta = \frac{1}{\sqrt{1 + (2\pi/\log(d))^2}}$$

Quarter amplitude damping, $d = 1/4$, gives the relative damping $\zeta = 0.22$, which is often considered too small. This is clearly seen in the examples above. The performance can be improved by the modification discussed below. In control loops where the major design objective is to quickly compensate for load disturbances, the high gain provided by the Ziegler-Nichols method is good. In these cases, large overshoots and oscillations during set point changes can be avoided by ramping the set point or performing the set point shift in several steps. In Section 2.4, another method to avoid large overshoots caused by set point changes was described.

Relations Between the Ziegler-Nichols Methods

The relations between the two methods can be seen by considering control of an integrator with a delay. Such a process has the transfer function

$$G(s) = \frac{b}{s} e^{-sT}$$

The step response parameters are $L = T$ and $a = bT$. The PID parameters obtained by the step response methods therefore become

$$K = \frac{1.2}{b \cdot T} \qquad T_i = 2T \qquad T_d = \frac{T}{2}$$

Design of PID Controllers

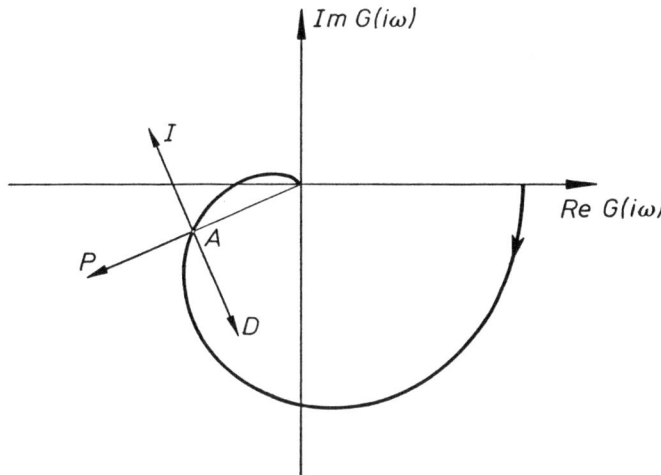

Figure 4.4
A Given Point on the Nyquist Curve May Be Moved to an Arbitrary Position in the G-plane by PI, PD, or PID Control. (Point A may be moved in the directions $G(i\omega)$, $G(i\omega)/i\omega$, and $i\omega G(i\omega)$ by changing the proportional, integral, and derivative gain, respectively)

The ultimate period of the system is $t_c = 4T$, and the ultimate gain is $k_c = \pi/2bT$. The PID parameters obtained by the frequency response methods are, therefore,

$$K = \frac{0.6\pi}{2bT} \approx \frac{0.94}{bT} \qquad T_i = 2T \qquad T_d = \frac{T}{2}$$

An Interpretation of the Ziegler-Nichols Frequency Domain Method

The Ziegler-Nichols frequency domain method will be interpreted in terms of moving points in the Nyquist diagram. The method starts with determination of the point $(-1/k_c, 0)$ where the Nyquist curve of the open-loop system intersects the negative real axis. With PI or PID control, it is possible to move a given point on the Nyquist curve to an arbitrary position

Design of PID Controllers

in the complex plane, as indicated in Figure 4.4. By changing the gain, it is possible to move the Nyquist curve in the direction of $G(i\omega)$, i.e., radially from the origin. Point A may be moved in the orthogonal direction by changing integral or derivative gain. It is thus possible to move a specified point to an arbitrary position, an idea that can be used to obtain design methods.

Let ω be the frequency that corresponds to A. The frequency response of the regulator at ω is

$$G_R(i\omega) = k \left[1 + \frac{1}{i\omega T_i} + i\omega T_d \right] = r_R e^{i\varphi_R}$$

with positive regulator parameters the angle φ_R is thus restricted to the range $-\pi/2 \leq \varphi_R \leq \pi/2$ where $\varphi_R = -\pi/2$ corresponds to pure integral control and $\varphi_R = \pi/2$ to pure derivative control.

Pure derivative control cannot be implemented (compare with Equation 2.7). The range of φ_R is therefore $-\pi/2 \leq \varphi_R \leq \varphi_0$ where φ_0 is about $\pi/3$ or 60°.

With the Ziegler-Nichols frequency response method it follows that

$$G_R(i\omega_c) = 0.6k_c \left[1 + i \left(\omega_c T_d - \frac{1}{\omega_c T_i} \right) \right]$$

$$= 0.6k_c \left[1 + i \left(\frac{2\pi}{8} - \frac{1}{\pi} \right) \right] = k_c(0.6 + 0.28i)$$

The Ziegler-Nichols frequency response method can thus be interpreted as finding regulator parameters so that the point where the Nyquist curve intersects the negative real axis is moved to $-0.6 - 0.28i$. This corresponds to a phase advance of 25° at ω_c.

A Modified Ziegler-Nichols Method

With the given interpretation, it is straightforward to generalize the Ziegler-Nichols frequency domain method. Other points of the Nyquist curve can be selected. They can also be moved to other positions. In this way it is possible to obtain design methods where the specifications are given in terms of amplitude margins or phase margins.

A general formulation is to start with a given point of the Nyquist curve of the process

$$G_p(i\omega) = r_p e^{i(\pi + \varphi_p)}$$

Design of PID Controllers

and to find a regulator so that this point is moved to

$$B = r_s e^{i(\pi+\varphi_s)}$$

An amplitude margin design corresponds to $\varphi_s = 0$ and $r_s = 1/A_m$ where A_m is the amplitude margin; a phase margin design corresponds to $r_s = 1$ and $\varphi_s = \varphi_m$ where φ_m is the specified phase margin; and the Ziegler-Nichols frequency domain method corresponds to $r_s = 0.66$ and $\varphi_s = 0.44$.

Writing the frequency response of the controller as

$$G_R(i\omega) = r_R e^{i\varphi_R}$$

we get

$$r_s e^{i(\pi+\varphi_s)} = r_p r_R e^{i(\pi+\varphi_p+\varphi_R)}$$

The controller should thus be chosen so that

$$\begin{cases} r_R = \dfrac{r_s}{r_p} \\ \varphi_R = \varphi_s = \varphi_p \end{cases}$$

Simple calculations give

$$k = \frac{r_s \cos(\varphi_s - \varphi_p)}{r_p}$$

$$\omega T_d - \frac{1}{\omega T_i} = \tan(\varphi_s - \varphi_p)$$

The gain k is uniquely given. However, only one equation determines the parameters T_i and T_d. An additional condition must thus be introduced to determine these parameters uniquely. A common method is to specify a constant relation between T_i and T_d, i.e.,

$$T_d = \alpha T_i$$

where α often is chosen as $\alpha = 0.25$. Straightforward calculations now give the parameters T_i and T_d.

$$T_d = \frac{1}{2\omega}\left[-\tan(\varphi_p - \varphi_p) + \sqrt{4\alpha + \tan^2(\varphi_p - \varphi_s)}\right]$$

$$T_i = \frac{1}{\alpha} T_i$$

Design of PID Controllers

For systems where the amplitude and the phase of the transfer function decreases monotonously, the choice $r_s = 0.5$ and $\varphi_s = \pi/4$ guarantees an amplitude margin of at least 2 and a phase margin of at least 45°.

Assuming that a Ziegler-Nichols experiment is used to determine a suitable point, we have $r_p = 1/k_c$ and $\varphi_p = 0$. The controller parameters are then given by $k = 0.35\, k_c$, $T_i = 0.77\, T_c$, and $T_d = 0.19 T_c$. This can be compared with the values given by the Ziegler-Nichols frequency response method.

The Ziegler-Nichols frequency response method and the modified Ziegler-Nichols method are based on the idea of moving one point on the Nyquist curve to a desired position. The terms phase margin and amplitude margin

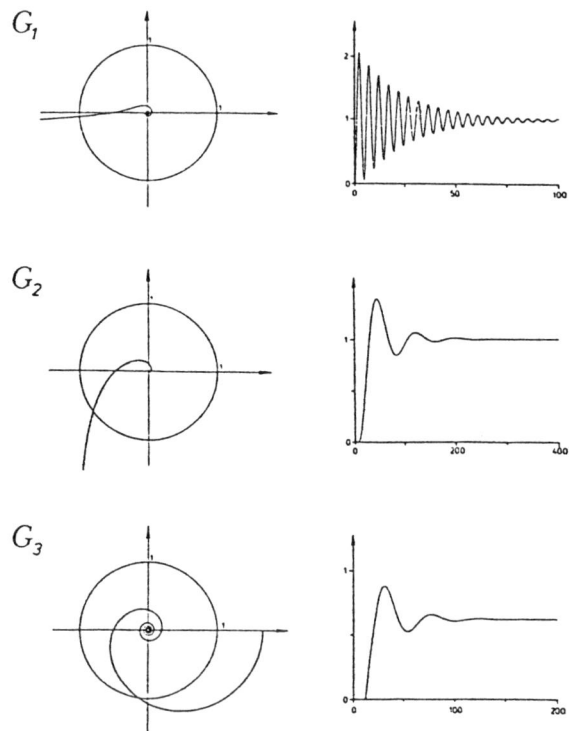

Figure 4.5
Nyquist Curves of Systems with Equal Amplitude Margin and Their Corresponding Closed-Loop Step Responses

Design of PID Controllers

also define one point on the Nyquist curve. In most cases these simple design rules are sufficient, but there are exceptions. Figure 4.5 shows the Nyquist curves of three systems having the same amplitude margin, $A_m = 2$. This means that all Nyquist curves pass through the point $z = -0.5$. Figure 4.6 shows the Nyquist curves of three systems having the same phase margin, $\varphi_m = 45°$. This means that all Nyquist curves pass through the point $z = -0.707 - 0.707i$. The corresponding step responses clearly demonstrate that the transient behavior of the control loop is also influenced by other points of the Nyquist curve. Design methods where several points on the Nyquist curve are determined are described below.

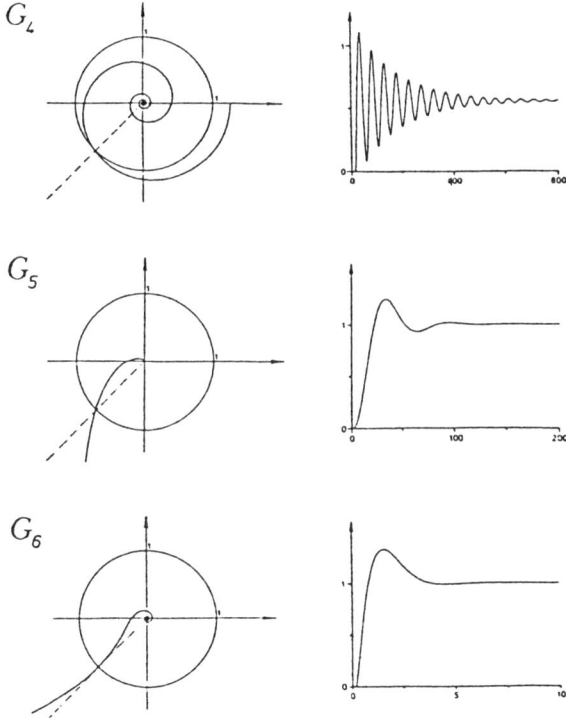

Figure 4.6
Nyquist Curves of Systems with Equal Phase Margin and Their Corresponding Closed-Loop Step Responses

Design of PID Controllers

4.3 DOMINANT POLE DESIGN

The Ziegler-Nichols methods discussed in the previous section were based on the knowledge of only one point on the Nyquist curve of the open-loop process dynamics. This section presents a design method that uses two points on the Nyquist curve. The method is based on a simple method of estimating the dominant poles of the closed-loop system from the open-loop transfer function. The notion of dominant poles is first discussed. The design method is then developed.

Dominant poles

Consider a closed-loop system obtained by negative feedback around a linear system with the transfer function $G(s)$ (see Figure 4.7). The transfer function of the closed-loop system from the command signal to the output is given by

$$G_c(s) = \frac{G(s)}{1 + G(s)}$$

Many properties of the closed-loop system can be deduced from the poles and the zeros of $G_c(s)$, which are the same as the zeros of $G(s)$ (i.e., the zeros of the plant and the controller). The closed-loop poles are the roots of the equation

$$1 + G(s) = 0$$

The pole-zero configurations of closed-loop systems may vary considerably. Many simple feedback loops will, however, have a configuration of the type shown in Figure 4.8 where the principal characteristics of the response are given by a complex pair of poles, p_1 and p_2, called the *dominant poles*. The response is also somewhat influenced by real poles and zeros p_3 and z_1, respectively. The position of z_1 and p_3 may be reserved. There may also be more poles and zeros far from the origin. Poles and zeros whose real parts are much smaller than the real part of the dominant poles have little influence on the transient response. Classical control was very much concerned with closed-loop systems having the pole-zero configuration shown in Figure 4.8.

Even if many closed-loop systems have a pole-zero configuration similar to the one shown in Figure 4.8, there are, however, exceptions. For instance, systems with mechanical resonances, which may have poles and zeros close to the imaginary axis, are generic examples of systems that do not fit the pole-zero pattern of the figure. Such systems are not treated in this section.

Design of PID Controllers

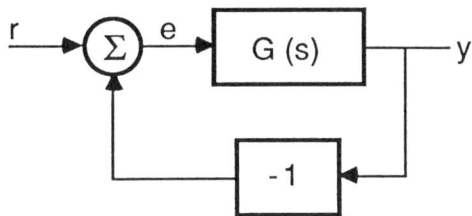

Figure 4.7
Block Diagram of a Simple Feedback System

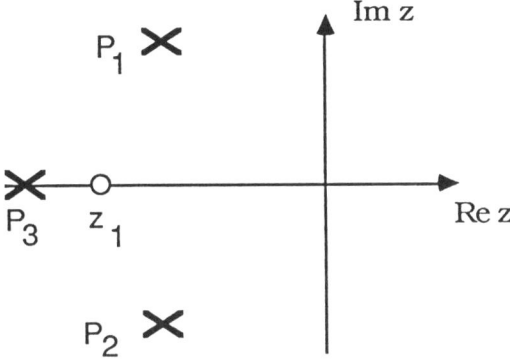

Figure 4.8
Pole-Zero Configuration of a Simple Feedback System

PI Control

The dominant pole design method will first be applied to PI control. Two closed-loop poles can be specified, since a PI controller has two adjustable parameters. Let the PI controller be parameterized as

$$G_R(s) = k + \frac{k_i}{s}$$

Design of PID Controllers

where k is the proportional gain and k_i is the integral gain. The parameters k and k_i will be determined so that the closed-loop system has poles at $s = p_1$ and $s = p_2$, where

$$p_1 = -\zeta\omega_0 + i\omega_0\sqrt{1-\zeta^2} = -\sigma + i\omega$$

$$p_2 = -\zeta\omega_0 - i\omega_0\sqrt{1-\zeta^2} = -\sigma - i\omega \qquad (4.2)$$

This implies

$$\begin{cases} 1 + \left[k + \dfrac{k_i}{p_1}\right] G_p(p_1) = 0 \\ 1 + \left[k + \dfrac{k_i}{p_2}\right] G_p(p_2) = 0 \end{cases}$$

where $G = G_R G_P$, and G_P is the transfer function of the process. The above equation is linear in k and k_i. It has a solution if $|G(p_1)| \neq 0$. The solution is

$$k(\omega_0) = -\frac{\sqrt{1-\zeta^2}\, A(\omega_0) + \zeta B(\omega_0)}{\sqrt{1-\zeta^2}\, [A(\omega_0)^2 + \zeta B(\omega_0)^2]}$$

$$k_i(\omega_0) = -\frac{\omega_0 B(\omega_0)}{\sqrt{1-\zeta^2}\, [A(\omega_0)^2 + B(\omega_0)^2]} \qquad (4.3)$$

where $A(\omega_0) = \text{Re } G_P(p_1)$ and $B(\omega_0) = \text{Im } G_P(p_1)$.

The parameter ω_0 can be viewed as a design parameter that determines the response speed. Small values of ω_0 give a slow system, and large values give a fast system. If the process dynamics are of first order, the closed-loop system only has two poles: p_1 and p_2. The design parameter ω_0 can then be chosen arbitrarily. For higher-order dynamics, the closed-loop system will, however, have more poles. For stable systems with poles on the real axis, these poles will have real parts that are greater than $-\zeta\omega_0$ for large ω_0. The condition that the poles p_1 and p_2 are dominating will thus give an admissible range of the design parameter ω_0. The upper bound of ω_0 can be determined from the condition that the largest pole on the real axis is at $s = -\alpha\omega_0$. For stable processes G_P, the function $A(\omega_0)$ is positive and $B(\omega_0)$ is small for small ω_0. It then follows that the proportional gain $k(\omega_0)$ is negative for small ω_0. Since it is normally desirable to have positive controller gains, a lower bound for the design parameter is given by the condition $k(\omega_{0I}) = 0$. The value ω_{0I} corresponds to pure integral control. Analogously, the value ω_{0P} corresponds to pure proportional control. An alternative to choosing

ω_0, based on pole dominance, is to select an ω_0 that gives the largest value of the integral gain. This gives values that are very close to those obtained from the condition of pole dominance. A physical interpretation of the condition will be given later in connection with the discussion of PID control.

PD Control

The dominant pole design can also be applied to PD control. Let the PD controller be parameterized as

$$G_R(s) = k + k_d s$$

and require that the closed-loop system has poles at p_1 and p_2 given by Equation 4.2. Calculations analogous to those for the PI controller give

$$k(\omega_0) = \frac{-\sqrt{1-\zeta^2}\, A(\omega_0) + \zeta B(\omega_0)}{\sqrt{1-\zeta^2}\, [A(\omega_0)^2 + B(\omega_0)^2]}$$

$$k_d(\omega_0) = \frac{B(\omega_0)}{\omega_0 \sqrt{1-\zeta^2}\, [A(\omega_0)^2 + B(\omega_0)^2]}$$

(4.4)

where $A(\omega_0) = Re\, G_P(p_1)$ and $B(\omega_0) = Im\, G_P(p_1)$. Notice that PI and PD control are complementary since Equation 4.4 gives $k_d(\omega_0) < 0$ for $\omega < \omega_{0p}$ and (4.3) gives $k_i(\omega_0) < 0$ for $\omega > \omega_{0p}$. The design parameter ω_0 is thus always larger for PD control than for PI control as can be expected. An upper bound for the design parameter for PD control is given by the condition $k(\omega_{0D}) = 0$, where parameter ω_{0D} corresponds to pure derivative control. A reasonable choice of the design parameter is the value that corresponds to the largest value of the proportional gain. Another alternative is to analyze the conditions for pole dominance.

PID Control

With PID control, it is possible to position three closed-loop poles. Let the transfer function of the PID regulator be parameterized as

$$G'_R(s) = k' + \frac{k'_i}{s} + k'_d s$$

where k' is the proportional gain, k'_i the integral gain, and k'_d the derivative gain. Two closed-loop poles will first be positioned according to Equation 4.2, as was done for PI control. Assume that the PI design problem is

Design of PID Controllers

already solved, i.e., that the functions $k(\omega_0)$ and $k_i(\omega_0)$ given by Equation 4.3 are known. The value of the regulator transfer function G'_R at $p_1 = -\sigma + i\omega$ is

$$G'_R(-\sigma + i\omega) = k' + \frac{k'_i}{-\sigma + i\omega} + k'_d(-\sigma + i\omega)$$

$$= k' - \frac{\sigma k'_i}{\omega_0^2} - \sigma k'_d + i\left[-\frac{\omega k'_i}{\omega_0^2} + \omega k'_d\right]$$

Requiring that this transfer function has the same value as the transfer function for PI control gives

$$k' - \frac{\sigma k'_i}{\omega_0^2} - \sigma k'_d = k - \frac{\sigma k_i}{\omega_0^2}$$

$$-\frac{\omega k'_i}{\omega_0^2} + \omega k'_d = -\frac{\omega k_i}{\omega_0^2}$$

Hence,

$$k'(\omega_0) = k(\omega_0) + 2\sigma k'_d = k(\omega_0) + 2\zeta\omega_0 k'_d$$

$$k'_i(\omega_0) = k_i(\omega_0) + \omega_0^2 k'_d \tag{4.5}$$

Thus, there is a two-parameter (ω_0, k'_d) family of gains for a PID regulator, which gives a closed-loop system with poles at $s = p_1$ and $s = p_2$. The parameter k'_d will now be determined so that the closed-loop system also has a pole at $s = -\omega_0$.
Hence,

$$1 + \left[k' - \frac{k'_i}{\omega_0} - k'_d\omega_0\right]G_P(-\omega_0) = 0$$

Inserting the expressions in Equation 4.5 for k' and k'_i gives

$$1 + \left[k - \frac{k_i}{\omega_0} + 2\sigma k'_d - 2\omega_0 k'_d\right]G_P(-\omega_0) = 0$$

If $G_P(-\omega_0) \neq 0$, this equation can be solved with respect to k'_d. The solution is

$$k'_d(\omega_0) = \frac{1 + [k(\omega_0) - k_i(\omega_0)/\omega_0]G_P(-\omega_0)}{2\omega_0(1 - \zeta)G_P(-\omega_0)} \tag{4.6}$$

Design of PID Controllers

Equations 4.5 and 4.6 define a one-parameter (ω_0) family of controller gains, which gives a closed-loop system with poles at $-\zeta\omega_0 \pm i\omega_0 \sqrt{1-\zeta^2}$ and $-\omega_0$. The parameter ω_0 may be viewed as a design parameter. Small values of ω_0 give a slow system, and large values a fast system. If there are no constraints on the signs of the regulator gains and the system dynamics are of second order, arbitrary values of design parameter ω_0 can be specified, since the closed-loop system has only three poles. For systems with higher-order dynamics, the condition that the chosen poles are dominating will give constraints on the design parameter.

Example 4.3—The properties of the dominant pole design method will now be illustrated. Consider a process with the transfer function given in Equation 4.1. A PI regulator that gives closed-loop poles with relative damping $\zeta = 0.707$ will first be designed. The smallest value of the design parameter that gives a nonnegative proportional gain is $\omega_{0I} = 0.62$. This corresponds to pure integral control, i.e., $k = 0$ and $k_i = 0.394$. For $\omega_0 = 2.23$, the closed-loop system has poles at $-1.58 \pm 1.58i$, -2.24, -20.6, and -100. The controller gains are $k = 1.62$ and $T_i = 0.70$. A comparison with Example 4.1 and Example 4.2 shows that the parameters obtained by the dominant pole design are significantly different from those obtained by the Ziegler-Nichols methods. For larger values of ω_0, the pole at -2.24 will move towards the right and become dominating. For sufficiently large ω_0, gain k_i becomes negative. Controller gains for some different values of ω_0 are shown in Table 4.3. The integral gain (k_i) has its largest value for $\omega_0 = 2.45$. The parameters are $k = 1.73$ and $T_i = 0.74$, which are close to the values obtained for pole dominance. The integral gain becomes zero for $\omega_0 = 3.72$. PD control can be used for larger values of ω_0. The controller parameters for PD control are also shown in Table 4.3. The proportional gain has its largest value for $\omega_0 = 11.0$, and becomes zero for $\omega_0 = 16.65$. With PID control, the design parameter can be increased significantly compared with PI control. Table 4.4 shows the parameters obtained for different ω_0. The closed-loop system will have a double pole at $s = -\omega_0$ for $\omega_0 = 7.16$. The regulator parameters are then $k = 11.9$, $T_i = 0.45$, and $T_d = 0.115$. To assess the different designs, first observe that the time to the peak of a step response is approximately $4.5/\omega_0$. The value of ω_0 can thus be used to determine the response time. The value of the integral gain (k_i) is also useful to assess the response to load disturbances. Consider a step in load disturbances. The control law is given by

$$u(t) = ke(t) + k_i \int^t e(s)ds + k_d \frac{de}{dt}$$

Design of PID Controllers

Table 4.3
PI and PD Regulators for Different Values of Design Parameter ω_0

ω_0	k	k_i	T_i	k_d	T_d
0.62	0	0.394			
1.00	0.51	0.887	0.574		
2.00	1.48	2.16	0.684		
2.20	1.60	2.29	0.700		
2.30	1.66	2.33	0.713		
2.40	1.71	2.35	0.728		
2.45	1.73	2.35	0.738		
2.50	1.76	2.35	0.748		
2.60	1.80	2.32	0.774		
3.00	1.90	1.96	0.972		
3.72	1.88	0	∞	0	0
4.00	2.24			0.080	0.036
6.00	5.09			0.598	0.118
8.00	7.73			1.02	0.132
10.0	9.43			1.33	0.141
11.0	9.72			1.45	0.150
12.0	9.52			1.55	0.163
14.0	7.36			1.67	0.226
16.0	2.33			1.68	0.322
16.65	0			1.66	∞

Table 4.4
PID Regulators for Different Values of Design Parameter ω_0

ω_0	k	k_i	T_i	k_d	T_d
2.00	1.15	1.70	-0.114		
2.24	1.62	2.31	0	0.704	0
3.00	3.32	4.98	0.336	0.668	0.101
4.00	5.78	10.0	0.705	0.587	0.122
5.00	8.20	16.1	0.996	0.510	0.121
6.00	10.3	22.1	1.21	0.466	0.118
7.00	11.8	26.2	1.35	0.500	0.115
7.16	11.9	26.5	1.37	0.452	0.115
7.50	12.2	26.8	1.40	0.454	0.115
8.00	12.3	26.1	1.42	0.473	0.115

Design of PID Controllers

Assume that the system is initially at rest. With a controller having integral action, the error and its derivative are then zero. Let the system be subject to a load disturbance. For systems with constant static gain, the load disturbance must be compensated with a change of the control signal Δu. This change is then given by

$$\Delta u = k_i \int_0^\infty e(s)ds$$

The error integral due to a load disturbance is then

$$\int_0^\infty e(s)ds = \frac{\Delta u}{k_i}$$

For a given load disturbance, the error integral is thus inversely proportional to k_i.

The properties of the different control laws can now be assessed. With pure integral control, design parameter ω_0 is 0.62 and k_i is 0.39. The peak time is then approximately 7.2s. With PI control, the design parameter can be chosen in the range $0.62 < \omega_0 < 2.5$. This means that the response time can be increased by a factor of 4 compared to pure integral control. The integral gain k_i can also be increased from 0.39 to 2.35, which means that the error integral for load disturbances can be reduced by a factor of 6 compared to pure integral control. Notice that the largest value of k_i is obtained for $\omega_0 = 2.45$.

With PD control, the design parameter can be chosen in the range $3.7 \leq \omega_0 \leq 11$, with proportional gains in the range $1.9 \leq k \leq 9.7$. The largest value of the loop gain with PD control is 9.7, which means that PD control can only be used if the largest steady-state error is less than 10%.

With PID control the design parameter ω_0 can be chosen in the range $0.6 \leq \omega_0 \leq 7.5$. The value $\omega_0 = 7.5$ gives a threefold increase of response time compared to PI control. The integral gain k_i can be increased from 2.35 for PI control to 26.8, which corresponds to an error integral for load disturbances that is more than 11 times smaller.

Responses to step changes in the set point and load are shown in Figure 4.9. The simulations support the results of the analysis.

The dominant pole design is useful since it gives predictable results. It has, however, the drawback that the transfer function must be known in the complex plane. Approximate methods, which require only the values of the Nyquist curve, are, therefore, developed below.

Design of PID Controllers

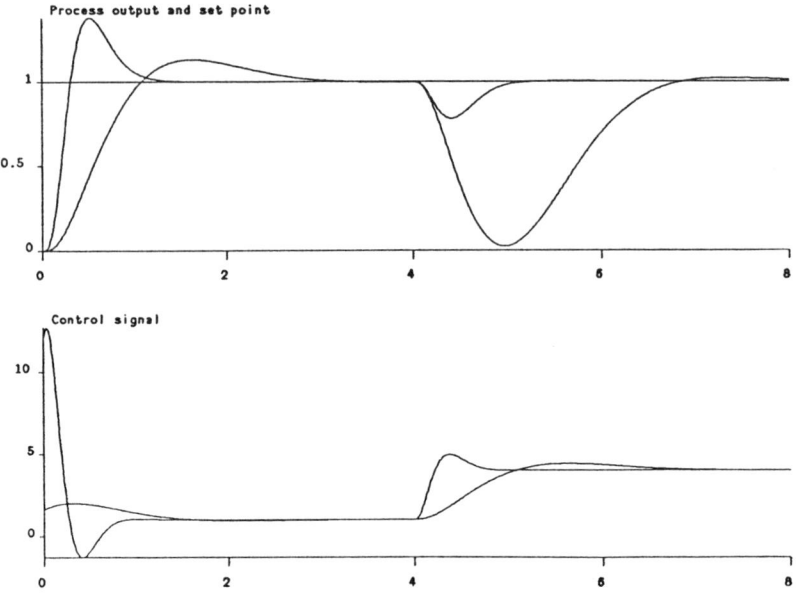

Figure 4.9
Step and Load Disturbance Responses of the Closed-Loop System in Example 4.3 Obtained with the Dominant Pole Design with $\omega_o = 7.16$

Approximate Determination of the Dominant Poles

The following is a simple method for estimating the dominant poles from knowledge of the Nyquist curve of the open-loop system. The closed-loop poles are given by the characteristic equation

$$G(s) + 1 = 0$$

A Taylor series expansion around $s = i\omega$ gives

$$0 = 1 + G(-\sigma + i\omega) = 1 + G(i\omega) + i\sigma G'(i\omega) + \ldots$$

where

$$G'(i\omega) = \frac{dG(i\omega)}{d\omega}$$

Design of PID Controllers

Neglecting terms of second and higher orders in σ, we find

$$1 + G(i\omega) + i\sigma G'(i\omega) = 0$$

Hence,

$$\sigma = i\frac{1 + G(i\omega)}{G'(i\omega)} \tag{4.7}$$

and both σ and ω of the dominant poles are determined. Notice that ω must be chosen so that σ becomes real. This analytic derivation shows that Equation 4.7 will give good results for small σ, i.e., when the dominant poles are close to the imaginary axis. The approximation will not hold if the function $G(s)$ has singularities inside a circle with the center in $i\omega$ and radius ω. This means that σ must be smaller than ω.

If the derivative is approximated by a difference between two close points on the Nyquist curve, the following expression for determining σ is obtained:

$$\frac{G(i\omega_2) - G(i\omega_1)}{\omega_2 - \omega_1} = i\frac{1+G(i\omega_2)}{\sigma} \tag{4.8}$$

By introducing a controller in the loop, the dominant poles may be moved to the desired new positions. The corresponding design problem may then be expressed in terms of the frequency (ω) and the relative damping (ζ) of the dominant poles.

To perform the design, it is assumed that the values of the open-loop transfer function at two neighboring frequencies, ω_1 and ω_2, are known, i.e.,

$$G_o(i\omega_1) = a_1 + ib_1$$
$$G_o(i\omega_2) = a_2 + ib_2$$

It is also assumed that frequencies ω_1 and ω_2 are close to the crossover frequency. The design is not restricted to any particular controller structure, and almost any controller with at least two adjustable parameters may be used. A PID controller of the form

$$G_R(s) = K\left[1 + \frac{1}{sT_i} + sT_d\right]$$

is chosen as an illustration. Furthermore, it is assumed that there is a given relation between the integration time (T_i) and the derivative time (T_d).

$$T_d = \alpha T_i \tag{4.9}$$

Design of PID Controllers

Hence,

$$G_R(s) = K\left[1 + \frac{1}{sT} + s\alpha T\right]$$

This regulator has two adjustable parameters: gain K, which moves the Nyquist curve radially from the origin, and time constant T, which twists the curve.

The design problem is then to determine a controller so that the transfer function of the compensated system has desired values at the two frequencies, i.e.,

$$\begin{aligned} G(i\omega_1) &= G_o(i\omega_1)\, G_R(i\omega_1) = c_1 + id_1 \\ G(i\omega_2) &= G_o(i\omega_2)\, G_R(i\omega_2) = c_2 + id_2 \end{aligned} \quad (4.10)$$

In the sequel, it is assumed that the desired frequency (ω) of the dominant poles is equal to ω_2. The following relation is then obtained from Equation 4.8:

$$\sigma = \frac{G(\omega_2) + 1}{G(i\omega_2) - G(i\omega_1)} \, i(\omega_2 - \omega_1)$$

The relative damping (ζ) is introduced by

$$\sigma = \frac{\zeta \omega_2}{\sqrt{1-\zeta^2}}$$

These two equations now give

$$\frac{G(i\omega_2) - G(i\omega_1)}{G(i\omega_2) + 1} = \frac{\sqrt{1-\zeta^2}}{\zeta} \cdot \frac{i(\omega_2 - \omega_1)}{\omega_2} \triangleq i\kappa$$

It follows from Equation 4.10 that

$$\frac{c_2 - c_1 + i(d_2 - d_1)}{c_2 + 1 + id_2} = i\kappa$$

This gives

$$\begin{cases} c_2 - c_1 + \kappa d_2 = 0 & (4.11) \\ d_2 - d_1 - \kappa(c_2 + 1) = 0 & (4.12) \end{cases}$$

These conditions determine parameters K and T of the PID regulator. Equation 4.11 gives a second-order equation for T, from which T is solved. Gain K is then obtained from Equation 4.12.

Design of PID Controllers

Example 4.4—Consider the system given by Equation 4.1. Two points on the Nyquist curve that are used for the design are given by

$$\begin{cases} G(8 \cdot i) = -0.0593 - i \cdot 0.0135 \\ G(10 \cdot i) = -0.0396 \end{cases}$$

Using these two values of $G(i\omega)$, the design method presented above can be applied. The following set of PID parameters is obtained for $\alpha = 0.25$, $\omega_2 = 10$, and $\zeta = 0.4$:

$K = 14.2 \qquad T_i = 0.407 \qquad T_d = 0.102$

Step and load disturbance responses of the closed-loop system are given in Figure 4.10.

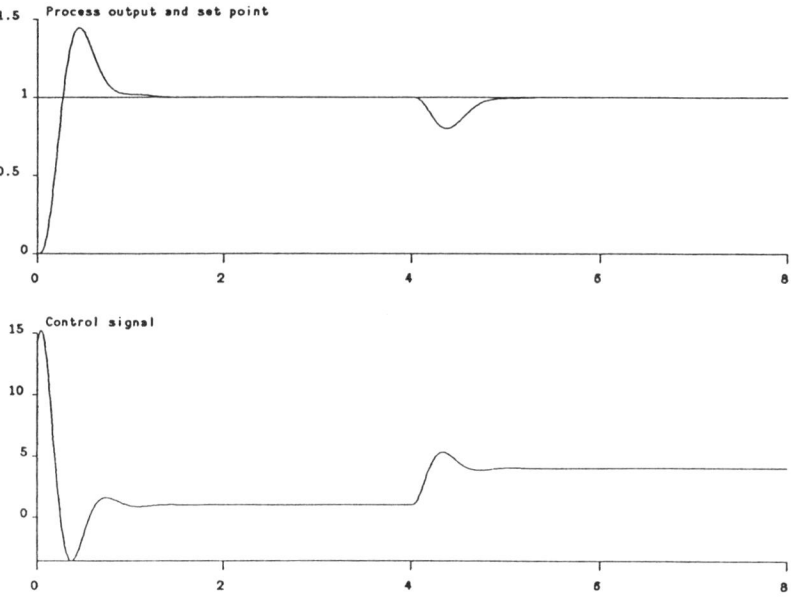

Figure 4.10
Step and Load Disturbance Responses of the Closed-Loop System in Example 4.4 Obtained with the Approximate Dominant Pole Design

Design of PID Controllers

A comparison of Figures 4.2 and 4.3 with Figure 4.10 shows that the responses obtained with the approximate dominant pole design are considerably better than those obtained by the Ziegler-Nichols methods. The price to be paid for the improved performance is that it is necessary to determine two points on the Nyquist curve of the open-loop system instead of one for the Ziegler-Nichols methods.

The parameters obtained by the approximate dominant pole design are quite similar to those obtained by the Ziegler-Nichols method. In the example, the gain is $K = 14.2$ versus 15 for the Ziegler-Nichols frequency response method. The other parameters are $T_i = 0.41$ (0.31) and $T_d = 0.10$ (0.08). The fact that the responses are different indicates that the parameter adjustment may be critical. This will be discussed further in Section 4.7.

Also notice that the design method is based on specification of only two parameters, σ and ω, the dominant poles. This implies that the gains of a PI or PD regulator are uniquely given. One extra condition has to be introduced to specify the three parameters of a PID controller, this condition being arbitrarily chosen as Equation 4.9.

4.4 FREQUENCY DOMAIN DESIGN

If several points on the Nyquist curve are known, many different design methods can be used. A common frequency domain approach attempts to find a compensator such that the magnitude of the closed-loop frequency response has unit gain at low frequencies and a resonance peak, M_P, which is less than a prescribed value. Such a design method is presented below.

M_P Values

Let $G = G_R G_P$ be the loop transfer function, i.e., the product of the transfer function of the controller and the process. The closed-loop transfer function is

$$G_s = \frac{G}{1 + G}$$

The curves in the G-plane where G_s has constant magnitude are given by

$$\left| \frac{G}{1 + G} \right| = M \qquad (4.13)$$

Design of PID Controllers

These are circles in the complex G-plane, called "M-circles". A few of the circles are shown in Figure 4.11.

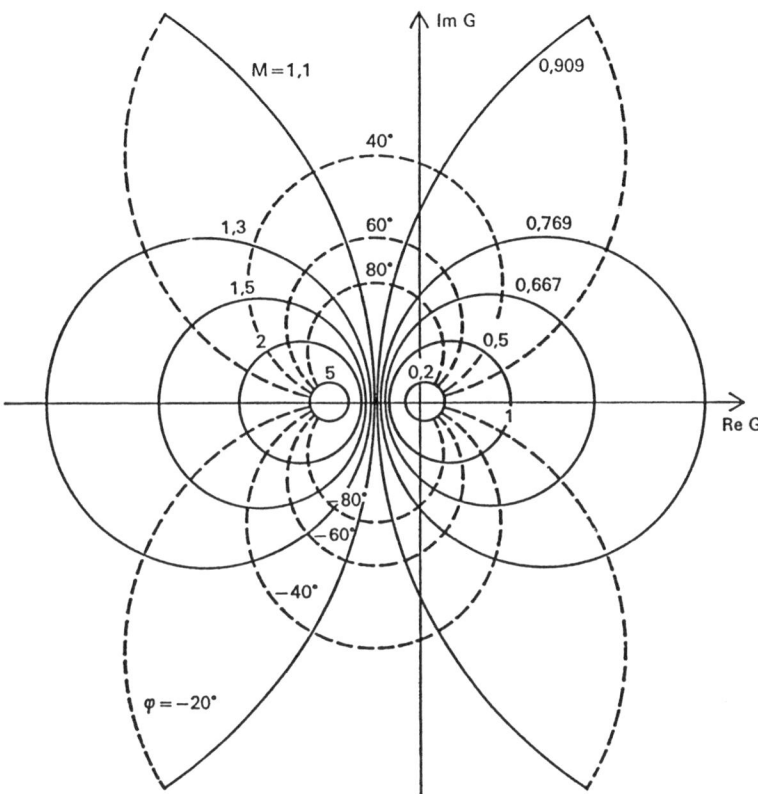

Figure 4.11
Complex G-plane with M-circles

Design of PID Controllers

The M_p value of a system is the largest value of M on its Nyquist curve. Notice that the Nyquist curve of a system is tangential to the M-circle, which corresponds to $M = M_p$. The M_p value can be related to other system characteristics and can be approximately computed from the relative damping (ζ) of the dominant poles in the following way:

$$M_p = \frac{1}{2\zeta\sqrt{(1-\zeta^2)}} \qquad \zeta \leq 1/\sqrt{2}$$

ζ is related to the full period damping as

$$d = e^{-\frac{2\pi\zeta}{\sqrt{1-\zeta^2}}}$$

Related values of M_p, ζ, and d are shown in Table 4.5, which also shows the radius (r) and the center (f) of the M-circles, given by

$$r = \frac{M}{M^2 - 1} \qquad f = \frac{M^2}{M^2 - 1}$$

Table 4.5.
Corresponding Values of M_p, Relative Damping (ζ), Absolute Damping (d), Radius (r) and Center (f) of the M-Circles.

M_p	ζ	d	r	f
1.1	0.54	0.018	5.24	5.76
1.2	0.47	0.034	2.72	3.27
1.3	0.42	0.052	1.88	2.45
1.4	0.39	0.071	1.46	2.04
1.5	0.36	0.091	1.20	1.80

Design Method

In the M-circle design method, the performance is specified by the M_p value, which is typically chosen in the range $M_p = 1.1$-1.5. The design rule is that the Nyquist curve of the compensated open-loop transfer function should avoid the interior of the circle associated with the specified M_p value and, instead, be tangent to it (see Figure 4.12).

Design of PID Controllers

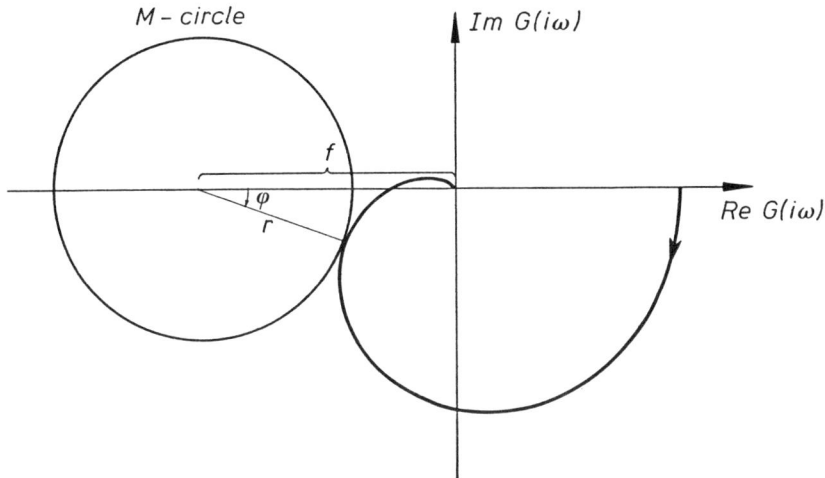

Figure 4.12
Graphical Illustration of the Design Procedure

The procedure can be described in some detail as follows. Let $G_p(i\omega)$ and $G_R(i\omega)$ denote the transfer functions of the process and the controller. Assume that the open-loop frequency response of the process is measured at frequency ω, i.e.,

$$G_p(i\omega) = a + b$$

Also assume that the derivative of G_p is measured at the same frequency. This can be done by measuring G_p at two neighboring frequencies. Hence,

$$G'_p(i\omega) = c + id$$

The transfer function of a PID controller is

$$G_R(i\omega) = K\left[1 + i\left(\omega T_d - \frac{1}{\omega T_i}\right)\right]$$

Hence,

$$G'_R = iK\left[T_d \; \frac{1}{\omega^2 T_i}\right]$$

Design of PID Controllers

Let the point where the compensated Nyquist curve touches the M_P circle be specified by angle φ (see Figure 4.12). This point is then given by the complex number:

$$A = -f + r\cos\varphi - ir\sin\varphi$$

The open-loop transfer function of the compensated system is

$$G = G_P G_R$$

Requiring that the compensated Nyquist curve goes through A gives

$$G_P(i\omega)\, G_R(i\omega) = -f + r\cos(\varphi) - ir\sin(\varphi) \tag{4.14}$$

Separating the real and imaginary parts of this equation gives

$$K\left[a - b\left(\omega T_d - \frac{1}{\omega T_i}\right)\right] = -f + r\cos\varphi$$

$$K\left[b + a\left(\omega T_d - \frac{1}{\omega T_i}\right)\right] = -r\sin\varphi$$

The condition that the compensated Nyquist curve is a tangent to the M_P circle at A can be expressed as

$$\arg G' = \arg(G'_P G_R + G_P G'_R) = \frac{\pi}{2} - \varphi \tag{4.15}$$

This equation implies that

$$\tan\varphi = \frac{c - d\left(\omega T_d - \dfrac{1}{\omega T_i}\right) - b\left(T_d + \dfrac{1}{\omega^2 T_i}\right)}{d + c\left(\omega T_d - \dfrac{1}{\omega T_i}\right) + b\left(T_d + \dfrac{1}{\omega^2 T_i}\right)}$$

We thus obtain three conditions: two for positioning the point and one for the slope. Since point A can be positioned anywhere on the chosen M circle, one extra degree of freedom can be chosen as angle φ in Figure 4.12, thus obtaining three conditions to determine four parameters (K, T_i, T_d, and φ). An auxiliary condition is obtained from

$$\omega T_i = \alpha \tag{4.16}$$

where α is a number in the range 3-6. This requirement implies that the integral action acts at a time scale that is compatible with the bandwidth (ω). With this additional requirement, the design procedure gives unique values of the PID parameters.

Design of PID Controllers

Validation

It is important to test the validity of a design based on simplified assumptions. First notice that the given procedure is based on local properties of the Nyquist curve; hence, there is no guarantee that the Nyquist curve will remain outside the M_p disc globally.

Although it is not possible to guarantee the properties of a design without access to detailed models or experiments, several quantities can be computed to obtain indications of the validity of a design.

The dimensionless quantity ωT_d can be interpreted as the normalized prediction horizon. This quantity should be small for the prediction to be good. To obtain a number, we can observe that a straight line prediction of a sinusoid can be made with a precision of 10% if

$$\omega T_d < 0.8 \tag{4.17}$$

Another quantity of importance is the ratio T_i/T_d. The numerator of the regulator transfer function has zeros at

$$s = \frac{1}{2T_d}\left[-1 \pm \sqrt{1-4T_d/T_i}\right]$$

If T_i/T_d is too small, the zeros will have poor damping. Since the closed-loop poles will migrate towards the zeros, we will thus require that $T_i > T_d$. This condition is automatically guaranteed by Equations 4.16 and 4.17.

A third condition is that the quantity

$$\gamma = \arctan\left(\omega T_d - \frac{1}{\omega T_i}\right)$$

(which represents the phase shift in the controller) is of reasonable magnitude, say less than $\pi/3$. This condition is also guaranteed by Equations 4.16 and 4.17.

It can thus be concluded that it is practical to impose the conditions of Equations 4.16 and 4.17 since this will automatically guarantee that other important conditions hold.

Design Variables

The design variables are the frequency (ω) and the M_p value. Although M_p values close to one will give systems with good damping, there are several drawbacks in choosing too small a value, because the associated M circle will then have a large radius, and it is then a greater risk that the

Design of PID Controllers

Nyquist curve will enter it at some other frequency. With a large radius of the M circle, the design will also be more sensitive. Reasonable values are therefore in the range of 1.3 to 1.5. According to Table 4.5, this corresponds to a relative damping around 0.4.

The frequency (ω) is also a critical variable. Experience has indicated that it is sensible to choose a frequency where the Nyquist curve of the process is in the third quadrant.

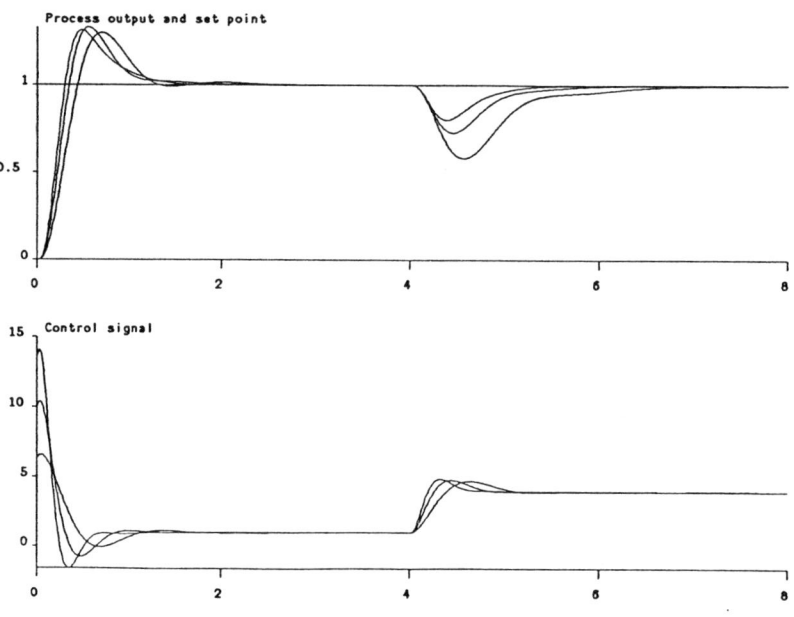

Figure 4.13
Step and Load Responses for the PID Controllers Obtained by the M circle Design Method. (The design parameters are $M_p = 1.3$ and $\omega T_i = 3$. The responses for $\omega = 4, 5,$ and 6 rad/s are shown.)

Design of PID Controllers

Example 4.5—The design procedure described above can be illustrated using the process model (Equation 4.1). The design parameters are chosen as $M_p = 1.3$ and $\omega T_i = 3$. Solving the design equations, the following controller parameters for $\omega = 4, 5, 6,$ and 7 are obtained:

ω	K	T_i	T_d	ωT_d	T_i/T_d
4	6.3	0.75	0.07	0.27	11.0
5	10.0	0.60	0.10	0.52	5.8
6	13.5	0.50	0.12	0.72	4.2
7	15.6	0.43	0.13	0.92	3.2

The PID controller obtained for $\omega = 4$ has phase lag. The value of ωT_d is a little too high for $\omega = 7$. This indicates that PID control can be used for bandwidths up to 6 rad/s but not higher with the chosen M_p value. Figure 4.13 shows the responses of the regulators obtained for $\omega = 4, 5,$ and 6 rad/s. If the M_p value is increased to 1.5, a valid design can be obtained for $\omega = 7$ rad/s. The parameters are $K = 14.8$, $T_i = 0.43$ and $T_d = 0.105$. For this design, $\omega T_d = 0.74$. A comparison with the previous results shows that the main effect of increasing M_p is that the derivation time decreases.

4.5 POLE PLACEMENT

The design methods presented previously in this chapter are all based on a limited knowledge of the process transfer function. Since the PID controller has only three design parameters, it cannot arbitrarily compensate more complicated process transfer functions. However, if the process is described by a low-order transfer function, a complete pole placement design can be performed, as described below.

PI Control of a First-Order System

Suppose that the process can be described by the following first-order model:

$$G_p = \frac{k_p}{1 + sT_1} \quad (4.18)$$

Design of PID Controllers

which has only two parameters, the process gain (k_p) and the time constant (T_1). By controlling this process with the PI controller,

$$G_R = K\left[1 + \frac{1}{sT_i}\right] \tag{4.19}$$

a second-order closed-loop system is obtained:

$$G_c = \frac{G_P G_R}{1 + G_P G_R} \tag{4.20}$$

The two closed-loop poles can be chosen arbitrarily by a suitable choice of the gain (K) and the integral time (T_i) of the controller. This is seen as follows. The poles are given by the characteristic equation, i.e., the equation

$$1 + G_P G_R = 0$$

The characteristic equation becomes

$$s^2 + s\left(\frac{1}{T_1} + \frac{k_p K}{T_1}\right) + \frac{k_p K}{T_1 T_i} = 0 \tag{4.21}$$

Now suppose that the desired closed-loop poles are characterized by their relative damping (ζ) and their frequency (ω). The desired characteristic equation then becomes

$$s^2 + 2\zeta\omega s + \omega^2 = 0 \tag{4.22}$$

Making the coefficients of these two characteristic equations equal gives two equations for determining K and T_i:

$$\begin{cases} \omega^2 = \dfrac{k_p K}{T_1 T_i} \\ \\ 2\zeta\omega = \dfrac{1 + k_p K}{T_1} \end{cases} \tag{4.23}$$

Hence, the following PI parameters are obtained:

$$\begin{cases} K = \dfrac{2\zeta\omega T_1 - 1}{k_p} \\ \\ T_i = \dfrac{2\zeta\omega T_1 - 1}{\omega^2 T_1} \end{cases} \tag{4.24}$$

Design of PID Controllers

Notice that in order to have positive controller gains it is necessary that the chosen bandwidth (ω) be larger than $1/(2\zeta T_1)$. Also notice that if ω is large the integration time T_i is given by

$$T_i = \frac{2\zeta}{\omega}$$

It is thus independent of the process dynamics for large ω. There is no formal upper bound to the bandwidth. However, a simplified model like Equation 4.18 will not hold for large frequencies. The upper bound on the bandwidth is therefore determined by the validity of the model.

PID Control of Second-Order Systems

Suppose that the process is characterized by the second-order model

$$G_p = \frac{k_p}{(1 + sT_1)(1 + sT_2)} \tag{4.25}$$

This model has three parameters. By using a PID controller, which also has three parameters, it is possible to arbitrarily place the three poles of the closed-loop system. The transfer function of the PID controller can be written as

$$G_R = \frac{K(1 + sT_i + s^2 T_i T_d)}{sT_i}$$

The characteristic equation of the closed-loop system becomes

$$s^3 + s^2 \left[\frac{1}{T_1} + \frac{1}{T_2} + \frac{k_p K T_d}{T_1 T_2} \right] + s \left[\frac{1}{T_1 T_2} + \frac{k_p K}{T_1 T_2} \right] + \frac{k_p K}{T_i T_1 T_2} = 0$$

A suitable closed-loop characteristic equation of a third-order system is

$$(s + \alpha \omega)(s^2 + 2\zeta \omega s + \omega^2) = 0 \tag{4.26}$$

which contains two dominant poles with relative damping (ζ) and frequency (ω), and a real pole located in $-\alpha \omega$. Identifying the coefficients in these two characteristic equations gives

Design of PID Controllers

$$\begin{cases} \dfrac{1}{T_1} + \dfrac{1}{T_2} + \dfrac{k_p K T_d}{T_1 T_2} = \omega(\alpha + 2\zeta) \\ \dfrac{1}{T_1 T_2} + \dfrac{k_p K}{T_1 T_2} = \omega^2(1 + 2\zeta\alpha) \\ \dfrac{k_p K}{T_i T_1 T_2} = \alpha\omega^3 \end{cases} \quad (4.27)$$

These three equations determine the PID parameters K, T_i, and T_d. The solution is

$$\begin{cases} K = \dfrac{T_1 T_2 \omega^2(1 + 2\zeta\alpha) - 1}{k_p} \\ T_i = \dfrac{T_1 T_2 \omega^2(1 + 2\zeta\alpha) - 1}{T_1 T_2 \alpha \omega^3} \\ T_d = \dfrac{T_1 T_2 \omega(\alpha + 2\zeta) - T_1 - T_2}{\omega^2 T_1 T_2 (1 + 2\zeta\alpha) - 1} \end{cases} \quad (4.28)$$

Notice that pure PI control is obtained for

$$\omega_c = \frac{T_1 + T_2}{(\alpha + 2\zeta) T_1 T_2}$$

Notice also that the choice of ω may be critical. The derivation time is negative for $\omega < \omega_c$. The frequency (ω_c) thus gives a lower bound to the bandwidth. Also notice that the gain increases rapidly with ω. The upper bound to the bandwidth is given by the validity of the simplified model (Equation 4.25).

Example 4.6—In this example, the model (Equation 4.1) is approximated with the second-order model

$$G_P(s) = \frac{1}{(1 + s)(1 + 0.26s)}$$

Here, the longest time constant of the model is kept, and the three shortest time constants are approximated with their sum. If $\zeta = 0.5$ and $\alpha = 1$ are chosen, the design calculation gives the following PID parameters:

Design of PID Controllers

$$\begin{cases} K = 0.52\omega^2 - 1 \\ T_i = \dfrac{0.52\omega^2 - 1}{0.26\omega^3} \\ T_d = \dfrac{0.52\omega - 1.26}{0.52\omega^2 - 1} \end{cases}$$

In this case, pure PI control is obtained for $\omega = 2.4$. The derivative gain becomes negative for lower bandwidths. The approximation neglects the load time constant 0.05. If the neglected dynamics are required to give a phase error of, at most, 0.3 rad (17 deg) at the bandwidth, $\omega < 6$ rad/s can be obtained. In Figure 4.14, the behavior of the control is demonstrated for $\omega = 4$, 5, and 6. It is straightforward to apply the direct design approach based on the simplified process models. The specification of the desired

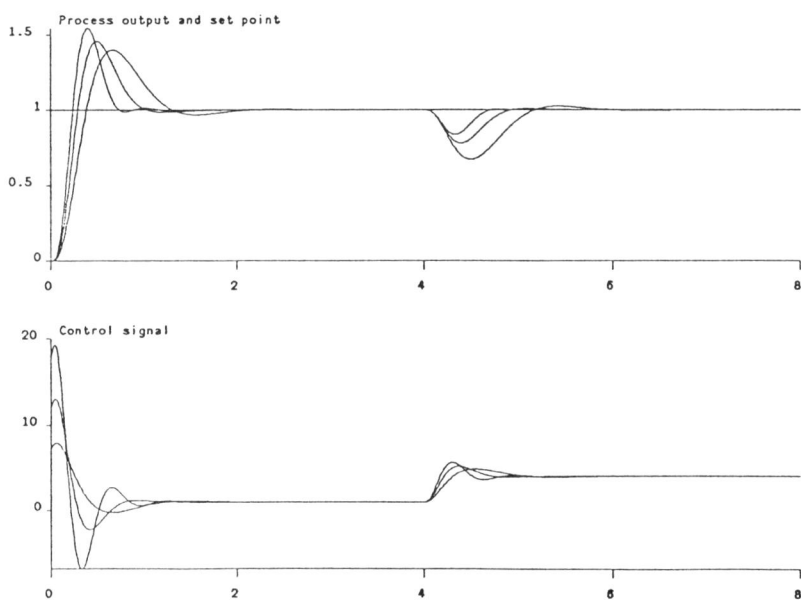

Figure 4.14
Step and Load Disturbance Responses of the Process (Equation 4.1) Controlled by a PID Controller Tuned According to Example 4.6 (The responses for $\omega = 4$, 5, and 6 are shown.)

Design of PID Controllers

closed-loop bandwidth is, however, crucial since the controller gain will increase rapidly with the specified bandwidth. It is crucial to know the frequency range where the model is valid. Alternatively, an upper bound to the controller gain can be used to limit the bandwidth. Notice the effect of changing the design frequency (ω). The system with $\omega = 6$ responds faster and has a smaller error when subjected to load disturbances. Simulations indicate that the design will not work well when ω is increased above 8.

Cancellation of Process Poles

A particular class of design methods is based on the idea of choosing the parameters of the controller so that the dominant process poles are canceled. These methods are quite popular because they are very simple and give a good response to set point changes. They will, however, often give poor response to load disturbances.

To explain the methods, consider the transfer functions of a PI controller:

$$G_R(s) = k\left[1 + \frac{1}{sT_i}\right] = \frac{k(1 + sT_i)}{sT_i}$$

and an ideal PID controller with error feedback:

$$G_R(s) = k\left[1 + \frac{1}{sT_i} + sT_d\right] = \frac{k[1 + sT_i + s^2 T_i T_d]}{sT_i}$$

One process pole can be canceled by a PI controller, and two process poles can be canceled by a PID controller. The response to load disturbances is poor for the designs based on cancellation because the dynamics corresponding to the canceled poles will appear in the response to the load disturbance. These modes will then recover in the same way as for the open-loop system. The same phenomena occur if the cancellation is not exact.

Example 4.7—PID design based on cancellation of process poles

Consider the system given by Equation 4.1. The system has the poles $p1 = -1$, $p2 = -1/0.2 = -5$, $p3 = -1/0.05 = -20$, and $p4 = -1/0.01 = -100$. Two of these poles can be canceled with a PID controller. Choosing the parameters T_i and T_d so that the slowest poles are canceled,

$$1 + sT_i + s^2 T_i T_d = (1 + s)(1 + 0.2s) = 1 + 1.2s + 0.2s^2$$

Design of PID Controllers

This gives $T_i = 1.2$ and $T_d = 0.167$. To find a suitable value of the controller gain, proceed as in the direct pole placement method in Example 4.6. The compensated transfer function becomes

$$G_R(s)G_P(s) = \frac{k}{sT_i(1 + 0.05s)(1 + 0.01s)} \approx \frac{k}{sT_i(1 + 0.06s)}$$

The characteristic equation of the closed-loop system is, therefore,

$$s(1 + 0.06s) + \frac{k}{T_i} = 0$$

or

$$s^2 + 16.7s + \frac{16.7k}{T_i} = 0$$

Identifying this with the characteristic equation

$$s^2 + 2\zeta\omega s + \omega^2 = 0$$

gives

$$k = \frac{4.2T_i}{\zeta^2} = \frac{5.0}{\zeta^2}$$

Choosing a relative damping $\zeta = 0.7$, then $k = 10$ and $\omega = 11.7$. Figure 4.15 shows the response of the closed-loop system obtained with these controller parameters. For comparison, the following results are obtained with a pole placement controller without cancellation. This controller has the parameters $k = 12$, $T_i = 0.37$, $T_d = 0.11$, $N = 10$ and $b = 0.35$. Notice the fast response to command signals and the poor response to load disturbances. Also notice the "spike" in the control signal, which depends on the fact that the derivative acts on the reference signal. The error due to a load disturbance decays with a time constant of $1s$, which corresponds to the cancelled mode $p1 = -1$ of the open-loop system. Because of the cancellation, the controller will not attempt to control this mode. This is clearly seen in the fact that the control signal settles much faster than the error at the load disturbance.

Although the designs based on cancellation of process poles are simple, they will not be discussed further because of their poor performance when subjected to load disturbances.

Design of PID Controllers

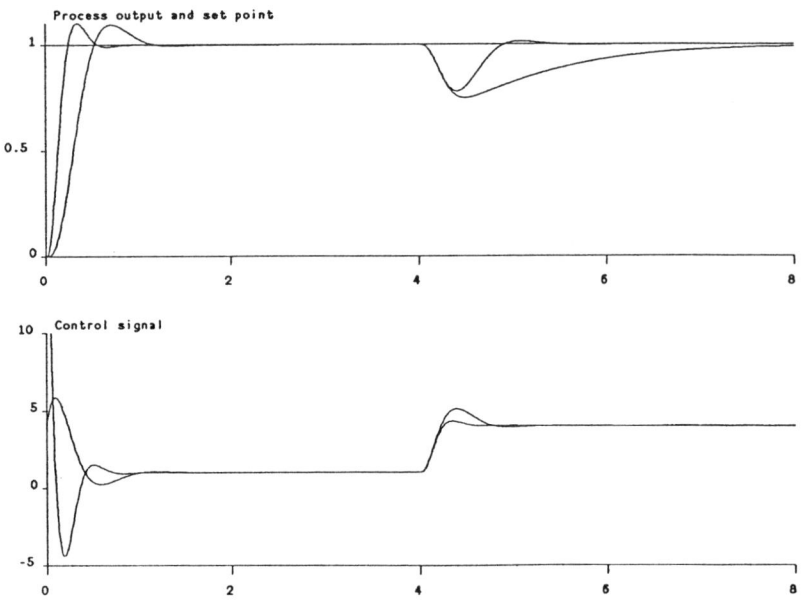

Figure 4.15
Simulation of PID Controller Based on Cancellation of Process Poles (For comparison, see the results of an equivalent design without cancellation shown by the thin lines.)

4.6 DISCRETE TIME POLE PLACEMENT

The examples have shown that PID controllers can be used for pole placement design when the process model is of low order. In the examples, continuous time models of the controller and the process have been used. It is also possible to use discrete time controllers for the pole placement design of discrete time process models, as shown below.

In Section 3.4, a discrete time process model was introduced using the z-transform instead of the Laplace transform used in continuous time models. Let the process be described by the transfer function

$$H_p(z) = \frac{B(z)}{A(z)} \tag{4.29}$$

Design of PID Controllers

Let $U(z)$ and $E(z)$ denote the z-transforms of the control signal, $u(t)$, and the error signal, $e(t)$. A general description of the controller is then

$$R(z)U(z) = S(z)E(z)$$

The transfer function of the controller can be written as

$$H_R(z) = \frac{U(z)}{E(z)} = \frac{S(z)}{R(z)} \tag{4.30}$$

The closed-loop transfer function is given by

$$H_C(z) = \frac{H_P H_R}{1 + H_P H_R} \tag{4.31}$$

and the characteristic equation therefore becomes

$$1 + H_P H_R = 0$$

Using Equations 4.29 and 4.30, the characteristic equation can also be written as

$$A(z)R(z) + B(z)S(z) = 0 \tag{4.32}$$

Now suppose that the process is of second order with the following transfer function polynomials:

$$A(z) = z^2 + a_1 z + a_2$$
$$B(z) = b_1 z + b_2$$

This structure of the process model captures many processes common in the process controller and is, for example, obtained by sampling the continuous time model (Equation 4.25) in the previous section. To ensure that the controller has integral action, the R-polynomial must be of the form

$$R(z) = (z-1)R_1(z)$$

The controller polynomials are given on the general forms

$$S(z) = s_0 z^2 + s_1 z + s_2$$
$$R(z) = (z - 1)(z + r_1)$$

Thus, the characteristic equation is obtained:

$$(z^2 + a_1 z + a_2)(z - 1)(z + r_1) + (b_1 z + b_2)(s_0 z^2 + s_1 z + s_2) = 0 \tag{4.33}$$

which is of fourth order. Assume that the desired closed-loop characteristic polynomial is given by

Design of PID Controllers

$$P(z) = (z - e^{-\alpha\omega h})^2(z^2 + p_1 z + p_2) \tag{4.34}$$

where

$$P_1 = -2 e^{-\zeta\omega h} \cos\left(\omega h \sqrt{1-\zeta^2}\right)$$

$$P_2 = e^{-2\zeta\omega h}$$

This corresponds to a fourth-order system having two dominant poles with relative damping (ζ) and frequency (ω), and two real poles located in $-\alpha\omega$.

The controller parameters can now be determined from the two descriptions of the characteristic equation, Equations 4.33 and 4.34. By comparing terms of equal power of z, parameters r_1, s_0, s_1, and s_2 can be determined, as illustrated in the following example. A detailed presentation of the discrete time design method is given in the book by Åström and Wittenmark (1984).

Example 4.8—In Example 4.6, the fourth-order model (Equation 4.1) was approximated by the second-order model:

$$G_P(s) = \frac{1}{(1+s)(1+0.26s)}$$

If this model is sampled with the sampling period $h = 0.1s$, the following discrete time model is obtained:

$$H_P(z) = \frac{0.0164z + 0.0140}{z^2 - 1.583z + 0.616}$$

If the design parameters are $\zeta = 0.5$, $\omega = 4$, and $\alpha = 1$, the desired characteristic polynomial becomes

$$(z - 0.670)^2(z^2 - 1.54z + 0.670$$

Comparing this characteristic polynomial with the one obtained according to Equation 4.33, the following set of controller parameters is obtained:

$$\begin{cases} r_1 = -0.407 \\ s_0 = 6.74 \\ s_1 = -9.89 \\ s_2 = 3.61 \end{cases}$$

Design of PID Controllers

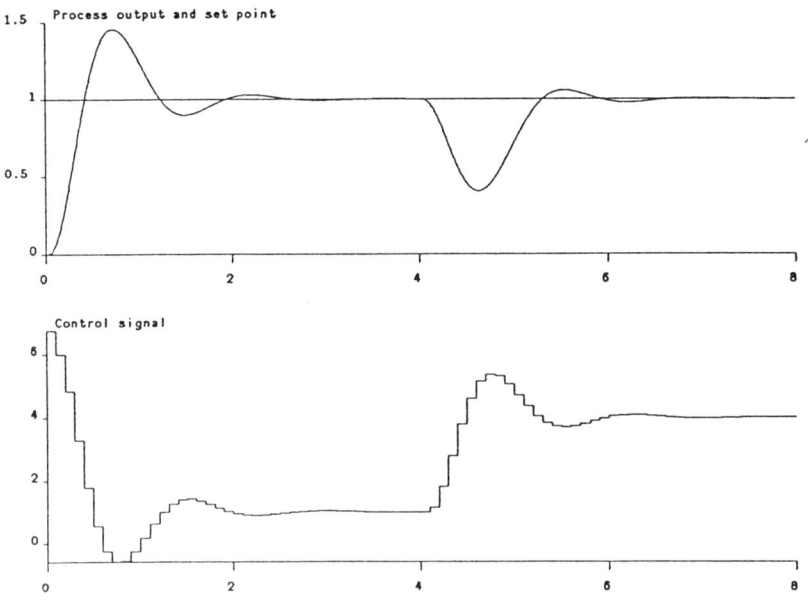

Figure 4.16
Step and Load Disturbance Responses of the Process (Equation 4.1) Controlled by a PID Controller Tuned According to Example 4.8

In Figure 4.16, the behavior of the control is demonstrated. Although the gain is fairly high (see the control signal), the response to the load disturbance is quite slow because of the low value of ω.

A drawback with direct digital design is that it is normally difficult to translate the controller to PID structure. The structure of the controller used in this section is such a case. On the other hand, this general form is useful when trying to cope with problems that are hard to solve with the standard PID controller. Such an example is dead time compensation, where a suitable controller can be derived just by introducing the dead time in the process model $H_p(z)$.

Design of PID Controllers

4.7 IMPROVEMENT OF SET POINT CONTROL

The controllers simulated in this chapter have responses to set point changes with excessive overshoot. Typical examples are given in Figures 4.2, 4.8 and 4.16. The reason for this is that the standard form of the PID controller with error feedback is used. The transfer function between the set point and the control signal of a PID controller is

$$G_{PID}(s) = K \frac{1 + sT_i}{sT_i}$$

The derivative part does not occur, since the derivation is performed on the process output only. The controller introduces a closed-loop zero at

$$s = -\frac{1}{T_i} \tag{4.35}$$

The influence of this zero was discussed in Section 2.4, where it was proposed to use a modified PID controller where only a fraction (b) of the reference signal is introduced in the proportional part. Such a controller is described by Equation 2.9, i.e.,

$$u = K\left[e_p + \frac{1}{T_i}\int_0^t e(s)ds + T_d \frac{de_d}{dt}\right] \tag{4.36}$$

where the error in the proportional part is

$$e_p = br - y$$

and the error in the derivative part is

$$e_d = -y$$

and the error in the integral part is

$$e = r - y$$

The modified controller has a zero at

$$s = -\frac{1}{bT_i}$$

which can be positioned properly by choosing the parameter b suitably. An estimate of the dominant closed-loop poles is necessary to do this. To avoid an excessive overshoot, parameter b should be chosen so that the zero is two to three times larger than the magnitude of the dominant poles. Estimates of the dominant poles are available for many of the design methods.

Design of PID Controllers

The Ziegler-Nichols Method

In the Ziegler-Nichols method, estimates of the dominant poles are obtained from the estimate of the closed-loop dominant period. This is listed as T_p in Tables 4.1 and 4.2. The design rule given above then gives b = 0.2 –0.3. Figure 4.17 shows simulations with the modified controller with b = 0, 0.2, 0.3, and 1.0. The figure shows clearly that the overshoot is reduced drastically when the modified algorithm is used. It also indicates that the design rule gives a reasonable value of b.

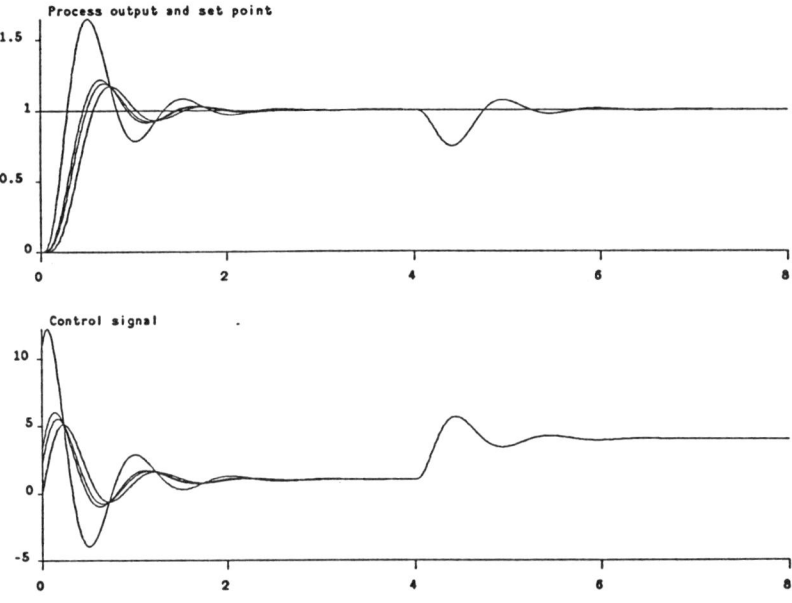

Figure 4.17
The Effect of Parameter b on the Step Response of a Closed-Loop System (The PID parameters are the same as in Figure 4.2.)

Design of PID Controllers

Direct and Dominant Pole Designs

In direct and dominant pole design methods, it is very easy to find good values for parameter b since these design methods deal directly with the dominant poles. Consider, for example, the direct design method used in Example 4.6 with $\omega = 6$ rad/s, which gives $T_i = 0.32$. The ordinary PID controller gives a zero at $s = 3.1$ rad/s, which is smaller than ω. To have the zero at $s = -12$, parameter b should be smaller than 0.26. To have the zero at $s = -18$, $b = 0.18$ should be chosen. Figure 4.18 shows a simulation of the modified PID controller. The figure shows clearly that the overshoot is reduced drastically when the modified algorithm is used. It also indicates that the rules for choosing parameter b are reasonable.

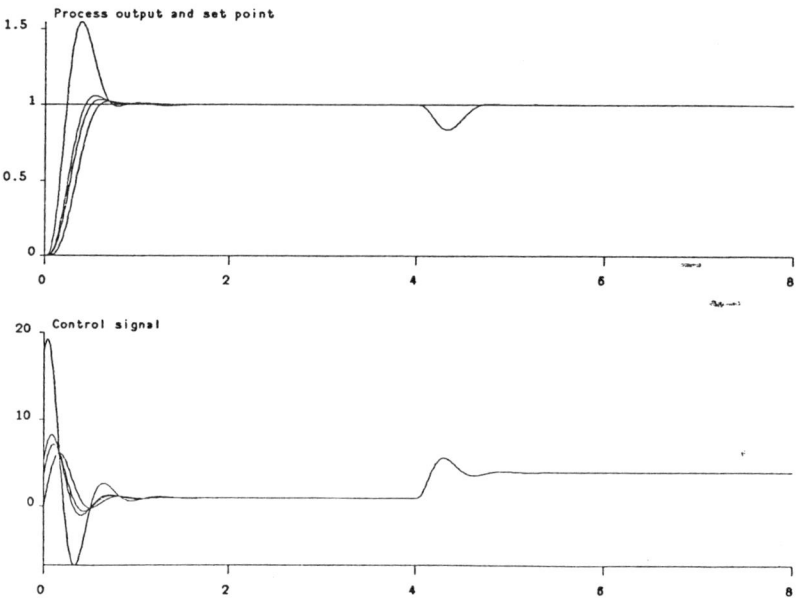

Figure 4.18
The Effect of Parameter b on the Step Response of a Closed-Loop System (The PID parameters are the same as in Figure 4.12 for $\omega = 6$ rad/s)

Design of PID Controllers

Even in the case of direct digital design it is possible to improve the responses to set point changes. In Section 4.6 the controller structure was given as

$$R(z)U(z) = S(z)E(z)$$

i.e., error feedback was used. If the extended structure

$$R(z)U(z) = -S(z)Y(z) + T(z)Y_R(z)$$

is used, the response to set point changes can be modeled by choosing polynomial $T(z)$ appropriately.

Conclusions

The results show conclusively that the responses to command signals are improved drastically by modifying the PID algorithm, as was discussed in Section 2.4.

4.8 COMPARISONS OF DESIGN METHODS

Although several methods have been given for designing PID controllers, all approaches have by no means been covered. There are many variations on the methods discussed herein, as well as a host of other techniques available in the literature. Instead of going on to describe more methods, it is a good idea to provide some perspective on the different methods. Before going into the details of the design methods, it can first be observed that control system design involves many different aspects, such as process dynamics, load disturbances, measurement noise, nonlinearities, and sensitivity. In this investigation, the focus has been on dynamics and set point changes, which is often adequate for the design of simple controllers.

Overview of the Approaches

The Ziegler Nichols Methods

These are simple approaches based on information on two parameters only, either L and a (which characterize the step response) or K_c and T_c (which characterize the frequency response).

Design of PID Controllers

The analysis leading to the dominant pole design indicates that it is not possible to give estimates of the closed-loop dominant poles from the knowledge of one point on the frequency response only. It can thus be concluded that there will always be a large uncertainty with design methods like the Ziegler-Nichols, which are based only on this information. All the other design methods discussed in this chapter use more information.

The Dominant Pole Design

The method is based on positioning two or three dominant poles. The method is based on knowledge of the plant transfer function at the dominant poles. Approximate methods based on knowledge of the frequency curves are also given. The dominant pole design method has one design parameter, namely the distance of the poles from the origin.

An interesting feature of the dominant pole design is that it gives ranges of the design parameter that are achievable with different controller types. This can be used to choose P, PI, PD, or PID control. We illustrate this point by an example.

Example 4.9—PI and PD Control of $(s+1)^{-3}$.

Consider a plant with the transfer function

$$G_P = \frac{1}{(s+1)^3}$$

Since the plant is of third order, it is clear that exact pole placement cannot be obtained with PI, PD, or PID control. First, consider PI control. Using the equation for the approximate dominant pole design, the following parameters are obtained:

$$k = \frac{\sigma(-4\omega^4+20\omega^2) + 3\omega^4 + 2\omega^2 - 1}{\omega^2 + 12\sigma^2 + 6\sigma + 1}$$

$$k_i = \frac{-\omega^6 + 2\omega^4 + 3\omega^2 - 12\sigma(\omega^4-\omega^2)}{\omega^2 + 12\sigma^2 + 6\sigma + 1}$$

where $k_i = k/T_i$. PD control gives instead

$$k = \frac{-2\sigma\omega^4 + 3\omega^4 + 16\sigma\omega^2 + 2\omega^2 - 6\sigma - 1}{\omega^2 + 6\sigma^2 + 6\sigma + 1}$$

$$k_d = \frac{\omega^4 + 12\sigma\omega^2 - 2\omega^2 - 12\sigma - 3}{\omega^2 + 6\sigma^2 + 6\sigma + 1}$$

Design of PID Controllers

where $k_d = kT_d$. The controllers will have positive gains only if the specifications on the dominant poles are restricted to certain values. Figure 4.19 shows the combinations of σ and ω that give positive gains for the PI and the PD controllers, respectively. The border lines are given by the pure P, I, and D controllers. Notice that the approximative formulas are only valid if $\sigma < \omega$. From this figure it is seen that the bandwidth ω cannot be chosen too high if only a PI controller is used.

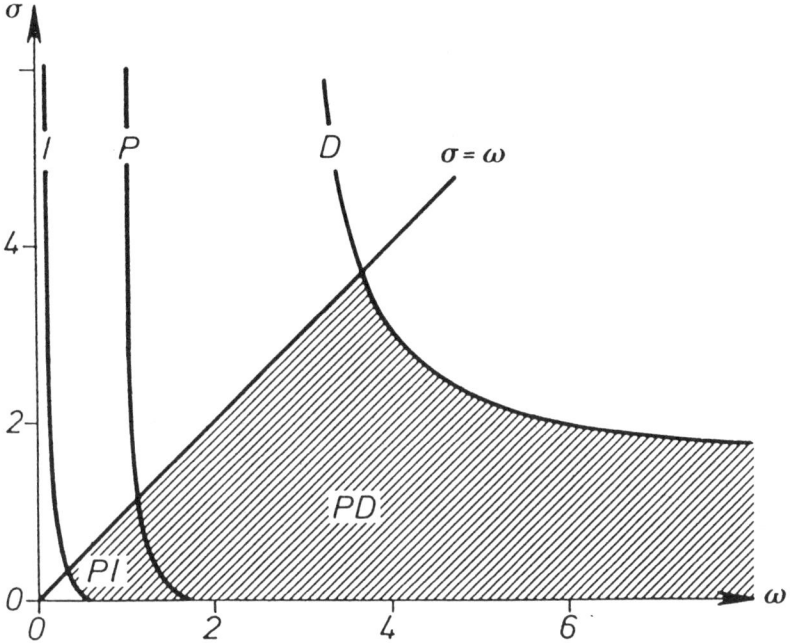

Figure 4.19
Regimes of Positive Gains for PI and PD Controllers

Design of PID Controllers

Simplified Frequency Domain Methods

This method is also based on knowledge of two *properly chosen* values of the open-loop frequency response of the system. The design method attempts to shape the closed-loop gain locally at the chosen frequencies.

When using the simplified frequency domain designs, it is clear that there are limitations on the shaping that can be done with a PID controller. It is thus necessary that the crossover frequency be chosen properly so that the loop can be shaped with a PID controller. It is also clear that the loop gain may behave badly at frequencies away from the chosen frequencies. This indicates that there will be problems with systems with resonances where the Nyquist curve twists and bends.

Pole Placement Methods

In the direct design methods, the dynamics are approximated by simplified models of first or second order, and the PID parameters are calculated from specifications on the desired closed-loop poles. The methods rely on making appropriate approximations and on the specifications being in harmony with these approximations.

Insight into the Problem

The direct design methods indicate superficially that any specification can be achieved. A closer inspection reveals, however, strict limitations. To obtain positive controller gains, it is necessary to choose the frequency (ω) sufficiently small (see Section 4.5). The formula for the controller gain also indicates that the gain will increase very rapidly with the chosen frequency. The frequency must also be chosen so low that the simplified model is valid well over ω. Experiments with continuous time and discrete time designs indicate that there is no large difference for small sampling periods. For longer sampling periods, the response to load disturbances will, however, be poorer for the discrete time algorithms because there will always be a time delay before the disturbance is captured.

Design of PID Controllers

Detailed Comparison

Table 4.6
Controller Parameters Obtained by the Different Design Methods

Method	K	T_i	T_d
Ziegler-Nichols step	10.9	0.32	0.08
Ziegler-Nichols frequency	15.0	0.31	0.08
Dominant pole design,			
$\omega = 5.3$	11.9	0.45	0.12
approximate method	14.2	0.41	0.10
M circle design			
$\omega = 4$	6.3	0.75	0.07
$\omega = 5$	10.0	0.60	0.10
$\omega = 6$	13.5	0.50	0.12
Direct pole placement			
$\omega = 4$	7.3	0.44	0.11
$\omega = 5$	12.0	0.37	0.11
$\omega = 6$	17.7	0.32	0.10
Direct pole placement with cancellation			
$\omega = 11.7$	10.0	1.20	0.17

Table 4.6 shows the parameters obtained when the different design methods are applied to the same problem. Several observations can be made from the table. First, with exception of the method based on cancellation, the controller parameters obtained by the different methods are similar. For example, the Ziegler-Nichols frequency domain method gives parameters that are quite close to the parameters obtained by the dominant pole design method. The main difference is that the gain of the Ziegler-Nichols method is too high and the derivation time is too low. Another interesting fact is that the Ziegler-Nichols method estimates the dominant frequency to be 12 rad/s, which is much too high. Also notice that the dominant pole design gives a value of the bandwidth ($\omega = \omega_0\sqrt{1-\zeta^2}$), but that ω has to be chosen by the designer for the direct pole placement.

Design of PID Controllers

Sensitivity

The comparison of the parameters obtained by the different methods indicates that the design may be quite sensitive to parameter variations. To investigate this, the parameters are perturbed in the Ziegler-Nichols frequency domain design. Figure 4.20 shows what happens when the derivation time is changed from T_d = 0.08 to 0.10 and 0.12. The figure indicates clearly that drastic improvements in the damping can be achieved by increasing the derivation time by 25%. Notice that the overshoot can be reduced drastically, as discussed in Section 4.7. This possibility was not used herein, because it is easier to see the improved damping with a large overshoot.

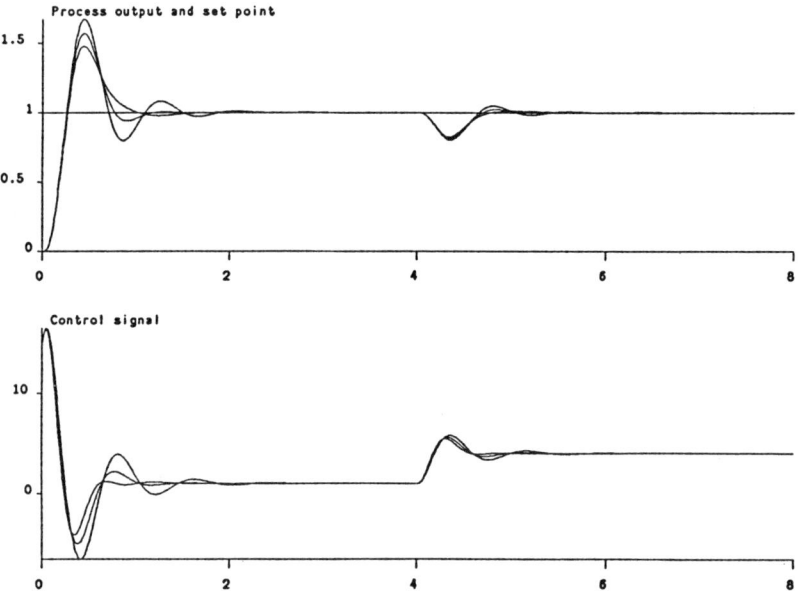

Figure 4.20
Effect of Changing the Derivation Time T_d in the Controller Obtained by the Ziegler-Nichols Frequency Domain Method

Design of PID Controllers

Figure 4.20 indicates that the system is sensitive, the reason being that the closed-loop bandwidth is quite high. It is a general rule that high bandwidth systems are sensitive. The fact that the bandwidth is high can be concluded from the comparison with the direct pole placement method. The analysis performed in Section 4.5 indicated that $\omega = 6$ rad/s was on the high side and that a more reasonable value is $\omega = 4$ rad/s. This is illustrated in Figure 4.21 and Figure 4.22, which illustrate the sensitivity of the direct designs for $\omega = 4$ rad/s and $\omega = 6$ rad/s to changes in the controller parameters. The derivation time (T_d) is changed by the same amount in both cases. Notice the drastic influence in particular on the closed-loop period and damping in Figure 4.22. The simulations strongly support reducing sensitivity by reducing the bandwidth.

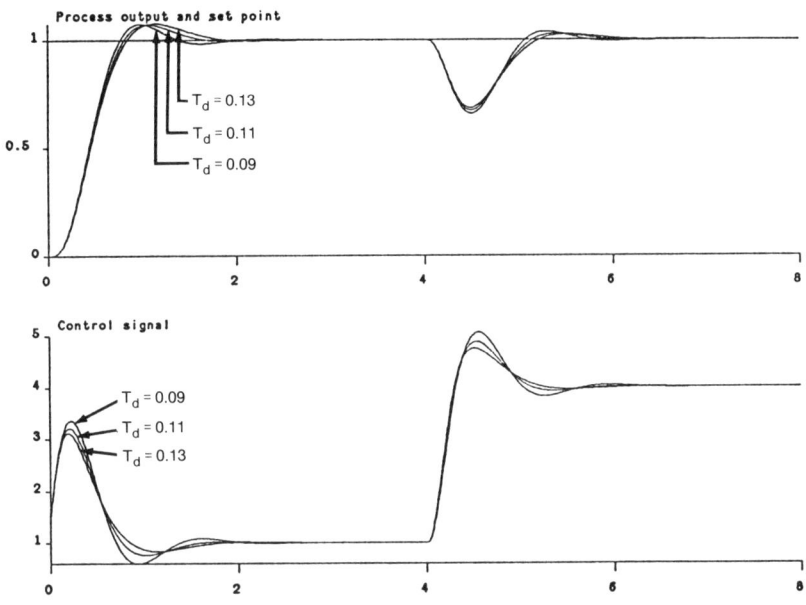

Figure 4.21
Effect of Changing Derivation Time T_d in the Controller Obtained by the Direct Pole Placement Design Method for $\omega = 4$ rad/s

Design of PID Controllers

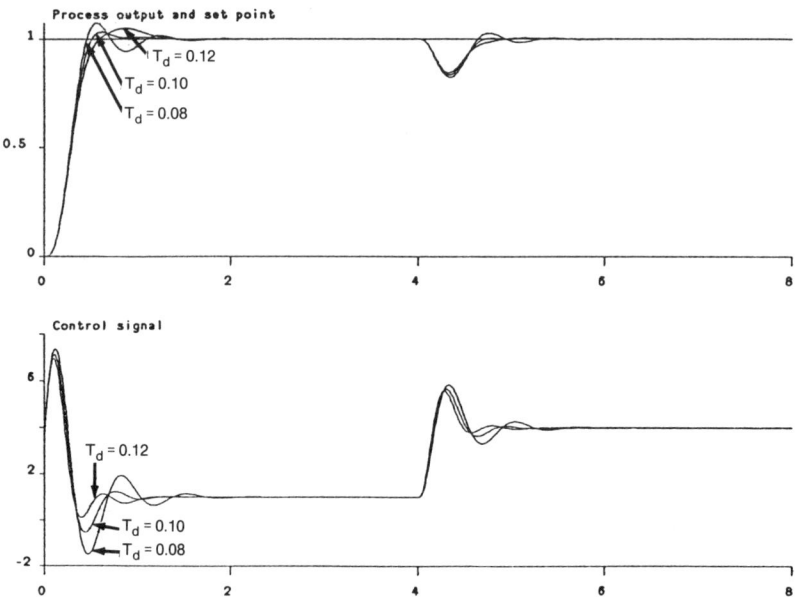

Figure 4.22
Effect of Changing Derivation Time T_d in the Controller Obtained by the Direct Pole Placement Design Method for $\omega = 6$ rad/s

Conclusions

For processes with simple dynamics, it has been demonstrated that it is possible to find design methods that give good results. Some insight into the properties of different design methods have been developed. In particular, the desired closed-loop bandwidth has been found to be a crucial specification; too high a bandwidth gives excessive gain and a sensitive system. With a controller like the PID, which has restricted complexity, it may not be possible to achieve the desired bandwidth. The choice of the bandwidth thus emerges as a key issue.

Design of PID Controllers

It would be highly desirable to have a procedure that would allow determination of an appropriate bandwidth automatically. The ultimate frequency is a good starting value, but the analysis of the Ziegler-Nichols tuning procedures indicates that this frequency may be too high.

Since a PID controller has a limited complexity, it is clear that arbitrarily large values of ω cannot be chosen. This is also clearly illustrated in the examples. It is also clear that the approach will always work for open-loop stable systems if ω is chosen sufficiently low.

The dominant pole design gives a suitable value of ω directly. The following guidelines are useful for design methods where ω has to be chosen. The open-loop crossover frequency (ω_c) can serve as a first approximation. The phase lead generated by a PID controller depends on the ratio $\alpha = T_d / T_i$ and the maximum derivative gain. With $\alpha = 0.25$, the largest lead is approximately 40°. This means that a proper phase margin may be obtained with $\omega = \omega_c$. To obtain a good transient response it is, however, also necessary that the slope $d \log | G(i\omega) | / d \log(\omega)$ is close to -1 at the crossover. Evaluation of the slope at the open-loop crossover frequency indicates whether the crossover frequency can be chosen as ω. There is again some margin. A PID controller can, for example, increase the slope by at most 0.4 when $\alpha = 0.25$. If the slope conditions can not be satisfied, a lower value of ω must be chosen.

Evaluating how rapidly the phase and the amplitude change also indicates whether the system is minimum phase. For a system with pure time delay, for example, the slope of the amplitude curve at the crossover is zero. To obtain a proper slope of the amplitude curve, it is then necessary to introduce PI control. The integration time should be chosen so that the integral action dominates at crossover. This means that derivative action is useless and that the time delay should give a phase shift of about 90° at the crossover.

4.9 CONCLUSIONS

In this chapter, several approaches to design PID controllers have been presented. The design methods are based on the different process models given in the previous chapter. The derivation of the process model and the design calculation are closely related. All design methods require a model of the process to be controlled. As has been seen, different design procedures are based on different process characterizations. Using a complex design procedure such as the full pole placement design requires a transfer function

Design of PID Controllers

description of the process with a high accuracy. The simple Ziegler-Nichols methods are based on limited process knowledge. If this kind of simple design procedure is used, there is no reason for making much effort in creating a detailed process model. Several design methods have been omitted in order to focus on those methods commonly used in the automatic tuning procedures. The process models described in Chapter 3 and the design methods presented in this chapter form the basis for the autotuning methods to be discussed in Chapter 5.

Autotuning

5

5.1 INTRODUCTION

By combining the methods for determination of process dynamics (described in Chapter 3) with the methods for computing the parameters of a PID controller (described in Chapter 4), methods for automatic tuning of PID controllers can be obtained. Practical controllers with such features have only recently appeared on the market. There are several reasons for this. The recent development of microelectronics has made it possible to incorporate the additional program code needed for the autotuning at a reasonable cost. The interest in autotuning at universities is also quite new. Most of the research effort has been devoted to the related but more difficult problem of adaptive control.

The chapter presents an overview of approaches to autotuning. Some commercial products are then examined: the Foxboro EXACT™ controller, which is based on transient response analysis; an autotuner based on relay feedback experiments and manufactured by Satt Control Instruments®; and two controllers based on parameter estimation and manufactured by Leeds & Northrup® and Turnbull Control Systems®. The last controllers also have a pretuning mode that is based on transient response analysis.

Autotuning

5.2 APPROACHES TO AUTOTUNING

A block diagram of a PID controller with automatic tuning is shown in Figure 5.1. When the tuning is performed, an extra loop is introduced. This loop consists of a process identification scheme and a design procedure that computes the PID parameters. An additional perturbation signal (v) may be added to ensure proper excitation of the process. The tuning can be performed either in closed loop or in open loop. The tuning procedure is characterized by:

(1) The underlying process model (see Chapter 3)
(2) The identification procedure (see Chapter 3)
(3) The design method (see Chapter 4)

The previous chapters thus provide the basis for discussing autotuning methods. Since there are several identification and design procedures, there are numerous ways to perform autotuning. Apart from the classification above, the autotuners can also be characterized by their operating modes. Tuning can be performed on operator demand or it can be initiated automatically.

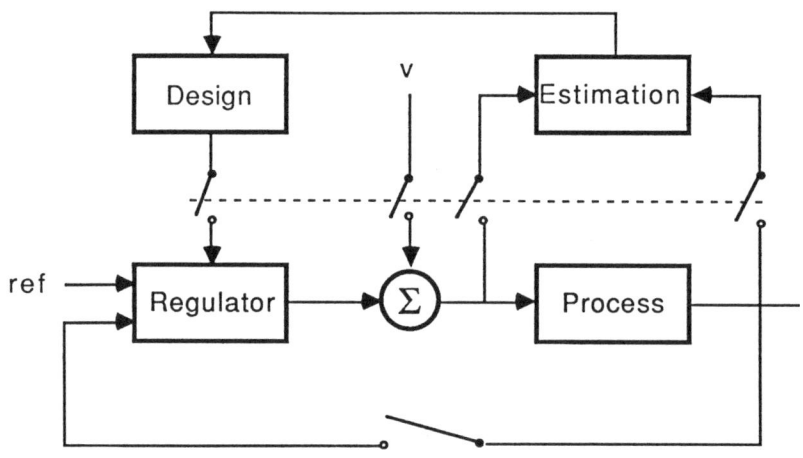

Figure 5.1
Block Diagram of a PID Controller with Automatic Tuning

Autotuning

Methods Based on Transient Responses

Autotuners can be based on open-loop or closed-loop transient response analysis. Methods for determining the transient response were discussed in Section 3.2. The most common methods are based on step or pulse responses, but there are also methods that can use a wide variety of transients.

Open-Loop Tuning

A simple process model can be obtained from an open-loop transient response experiment. A step or a pulse is injected at the process input, and the response is measured. To perform such an experiment, the process must be stable. If a pulse test is used, the process may include an integrator. It is necessary that the process be in equilibrium when the experiment is begun. There are, in principle, only one or two parameters that must be set *a priori:* the amplitude and the duration of a pulse or the amplitude of a step. These parameters must be chosen so that the response is well visible above the noise level, but not so large that the process nonlinearities become significant.

Many methods can be used to extract process characteristics from a transient response experiment. Periods of oscillation and damping can be determined from simple zero crossing and peak detectors (see Section 3.2). The method of moments discussed in the same section can also be used to determine process models directly. More elaborate signal processing methods, such as parameter estimation or the fast Fourier transform, can also be used.

The transient response methods are often used as a pretuning mode in more complicated tuning devices. The main advantage of the methods (they require little prior knowledge) is then exploited. The drawback with the transient response methods is that they are sensitive to disturbances. This drawback is less important if they are used only in the pretuning phase.

Closed-Loop Tuning

Automatic tuning based on transient response identification can also be performed on line. The steps or pulses are then introduced in the reference signal. They can be introduced by the controller for identification purposes. There are also autotuners that do not introduce any transient disturbances. Instead, large disturbances caused by set point changes or load disturbances

Autotuning

are detected and the closed-loop responses are considered. The Foxboro EXACT controller, which is described in the next section, is an example of such a scheme.

Since a proper closed-loop transient response is the goal for the design, it is appealing to base the tuning methods directly on the properties of the closed-loop transient responses. It is easy to give design specifications in terms of the closed-loop transient response, e.g., damping, overshoot, closed-loop time constants, etc. The drawback is that the relation between these specifications and the PID parameters is normally quite involved. Heuristics and logic are therefore required for these kinds of tuning devices.

Methods Based on Frequency Responses

Since frequency response methods can also be used to determine the process dynamics, as was discussed in Section 3.3, autotuners can be based on this methodology.

Use of the Relay Method

In traditional frequency response methods, the transfer function of a process is determined by measuring the steady-state response to a sinusoidal input. A difficulty with this method is that appropriate frequencies of the input signal must be specified. A special method where an appropriate frequency of the input signal is generated automatically was described in Section 3.3. The idea was simply to introduce a nonlinear feedback of the relay type so that there would be a limit cycle oscillation. With an ideal relay the method gives an input signal to the process whose period is close to the crossover frequency of the open-loop system.

A block diagram of an autotuner based on the relay method is shown in Figure 5.2. Notice that there is a switch that selects either relay feedback or ordinary PID feedback. When it is desired to tune the system, the PID function is disconnected and the system is connected to relay control. The system then starts to oscillate. The period and the amplitude of the oscillation are determined when steady-state oscillation is obtained. This gives the ultimate period and the ultimate gain. The parameters of a PID controller can then be determined from these values, e.g., using the Ziegler-Nichols frequency response method. The PID controller is then automatically switched in again, and the control is executed with the new PID parameters.

Autotuning

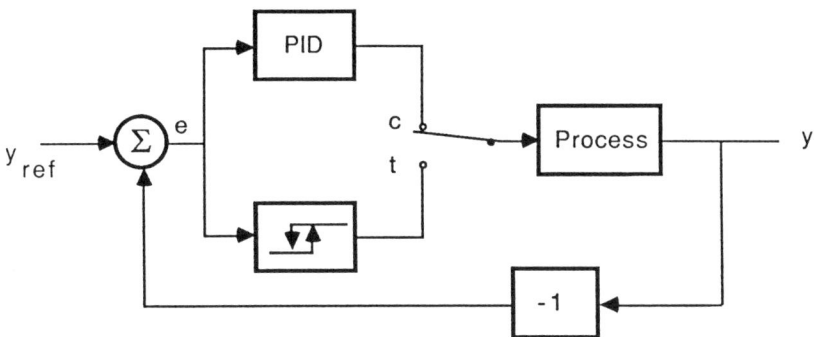

Figure 5.2
Block Diagram of an Autotuner Based on the Relay Method for System Identification (The system operates as a relay controller in the tuning mode (t) and as an ordinary PID controller in the control mode (c).)

This tuning device has one parameter that must be specified in advance, namely, the initial amplitude of the relay. A feedback loop from measurement of the amplitude of the oscillation to the relay amplitude can be used to ensure that the output is within reasonable bounds during the oscillation. It is also useful to introduce hysteresis in the relay. This reduces the effects of measurement noise and also increases the period of the oscillation. With hysteresis there is an additional parameter, which can, however, be set automatically based on a determination of the measurement noise level. Notice that there is no need to know time scales *a priori* since the crossover frequency is determined automatically.

In this method, an oscillation with suitable frequency is generated by a static nonlinearity. Even the order of the time constant of the process can be unknown. Therefore, this method is not only suitable as a tuning device; it can also be used in a pretuning phase in other tuning procedures where the time constant of the system has to be known, e.g., to decide a suitable sampling period.

This tuning method can also be modified to identify several points on the Nyquist curve. This can be accomplished by making several experiments with different values of the amplitude and the hysteresis of the relay. A filter with known characteristics can also be introduced in the loop to identify other points on the Nyquist curve. If two points are known, design methods based on two points can be used, e.g., the dominant pole design method or the M_p-circle design.

Autotuning

On-Line Methods

Frequency response analysis can also be used for on-line tuning of PID controllers. By introducing bandpass filters, the signal content at different frequencies can be investigated. From this knowledge, a process model given in terms of points on the Nyquist curve can be identified on line. A crucial choice in this autotuner is the choice of frequencies in the bandpass filters. This choice can perhaps be simplified by using the tuning procedure described above in a pretuning phase.

Methods Based on Parametric Models

Perhaps the most common tuning procedure is to use a recursive parameter estimation to determine a low-order discrete time model of the process. The parameters of the low-order model obtained are then used in some design scheme to obtain the controller parameters. An autotuner of this type can also be operated as an adaptive regulator that will change the regulator parameters continuously. Autotuners based on this idea, therefore, often have an option for continuous adaptation.

A drawback with autotuners of this type is that they require significant prior information. A sampling period for the identification procedure must be specified; it should be related to the time constants of the closed-loop system. Since the identification is performed on line, a controller that at least manages to stabilize the system is required. Therefore, the available products based on this identification procedure often have a pretuning phase, which can be based on the methods presented in the previous sections.

5.3 THE FOXBORO EXACT™

This controller is based on analysis of the transient response of the closed-loop system to set point changes or load disturbances and traditional tuning methods in the Ziegler-Nichols spirit. The idea behind it is a pattern recognition approach described in Bristol (1977). Some details about the actual implementation are given in Kraus and Myron (1984) and Kraus (1984).

Autotuning

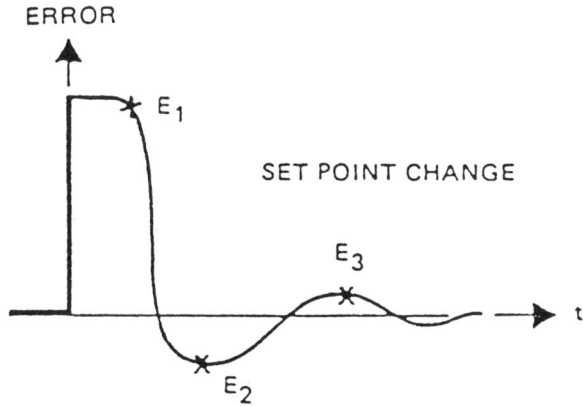

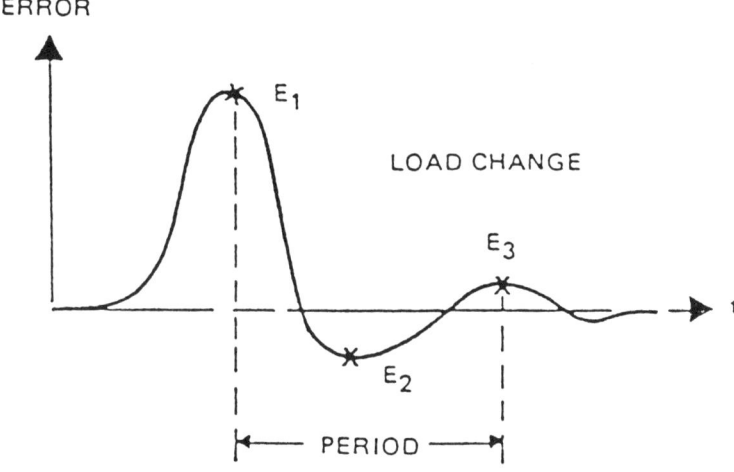

Figure 5.3
Typical Response of Control Error to Step or Impulse Disturbance

Autotuning

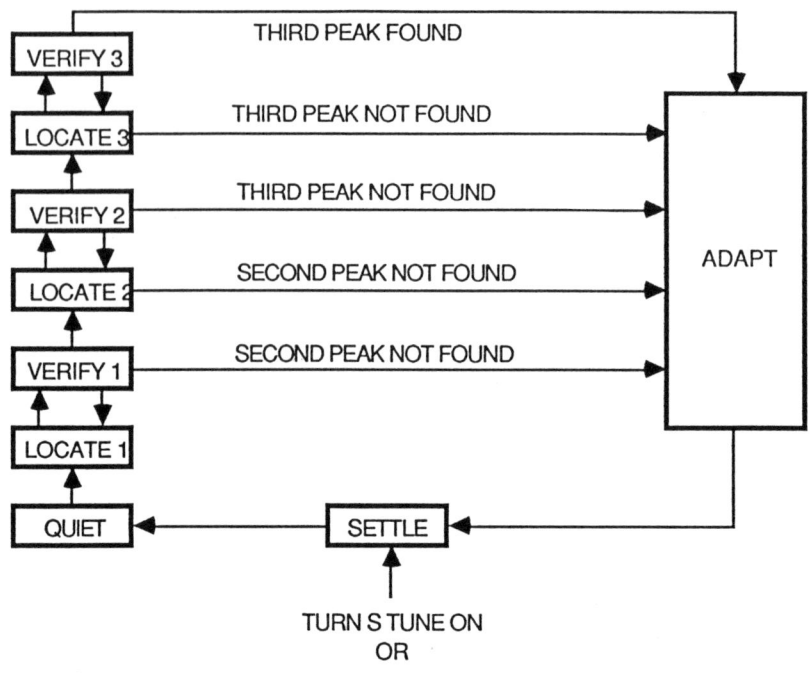

Figure 5.4
Heuristic Logic Used in the EXACT System to Determine the Characteristics of a Closed-Loop Transient

Process Modeling

The Foxboro system is based on the determination of dynamic characteristics from a transient, which results in a sufficiently large error. If the controller parameters are reasonable, a transient error response of the type shown in Figure 5.3 is obtained. Heuristic logic is used to detect that a proper disturbance has occurred and to detect the peaks e_1, e_2, and e_3 and the period T_p. The heuristic logic used is outlined in Figure 5.4. The estimation process is simple. It is, however, based on the assumption that the

Autotuning

disturbances are steps or short pulses. The algorithm can give wrong estimates if the disturbances are two short pulses, because T_p will then be estimated as the distance between the pulses. It also requires that the loop is closed with a controller that gives a reasonably stable response. A pretuning facility is therefore introduced to obtain the required information.

Control Design

The control design is based on specifications of damping, overshoot, and the ratios T_i/T_p and T_d/T_p, where T_i is the integration time, T_d the derivative time, and T_p the period of oscillation. The damping is defined as

$$d = \frac{e_3 - e_2}{e_1 - e_2}$$

and the overshoot as

$$z = \frac{e_2}{e_1}$$

In typical cases it is required that both d and z are less than 0.3. The Ziegler-Nichols tuning rule (Equation 4.2) gives

$$\frac{T_i}{T_p} = \frac{0.5}{0.85} \quad \text{and} \quad \frac{T_d}{T_p} = \frac{0.12}{0.85}$$

These ratios have been modified based on empirical studies in the Foxboro EXACT controller. They are computed from the dead time and the dominant process time constants. Smaller values are chosen for processes with dominant dead time, and larger values are selected for processes with a dominant lag. The details are not published.

The default values of the maximum overshoot and the maximum damping are 0.5 and 0.3, respectively. Both these parameters can be specified by the operator. The overshoot should be in the range [0,1] and the damping should be in the interval [0.1,1].

Prior Information and Pretuning

The tuning procedure requires prior information of the controller parameters K, T_i, and T_d. It also requires information of the time scale of the process. This is used to determine the maximum time the heuristic logic waits for the second peak. Some measure of the process noise is also needed

Autotuning

to determine that a disturbance has occurred and to set the tolerances in the heuristic logic. There are also some parameters that may be set optionally: damping (d), overshoot (z), the derivative factor, and bounds on the controller parameters. In the controller are also a number of safeguards; for example, the parameters are not allowed to change too much in a given adjustment.

There is also a pretune mode that can be used to initialize the controller parameters. This mode is activated as follows. The controller is first set in manual mode. The pretuning mode is then activated, which generates a step response. The step size has to be chosen large enough to cause a change in the measured value of at least 2.5%. The default value of the step is 8%. A dead time and a time constant are determined from the step response. The dead time estimate is used to determine the integral time, the derivative time, and the maximum wait time. The controller gain is determined from the dead time and the time constant. Finally, the noise level is determined from peak-to-peak measures at steady state. The pretune mode should be invoked only when the process is in steady state.

Operator interface

Figure 5.5 shows the front panel of the controller. All settings are made using keys on the front panel of the controller; no terminal is used. Three keys have fixed functions, while the other five are multifunction keys. All parameters available to the operator are given in an hierarchical structure diagram. The multifunction keys are used to move in different directions in the diagram and to change parameter values. Some parameters of particular importance for the tuning function are listed below.

NB   Noise band. The tuning is activated when the error exceeds twice the noise band.

WMAX   Maximum wait time. This is the maximum time that the algorithm waits for a second peak.

DMP   Maximum allowed damping. It can be chosen in the range [0.1,1]; the default value is 0.3.

OVR   Maximum allowed overshoot. It can be chosen in the interval [0,1]; the default value is 0.5.

CLM   Change limit. Limits the controller parameters within a certain range. Expressed as a fraction and a multiple of the initial P, I, and D settings. Default value is 10.

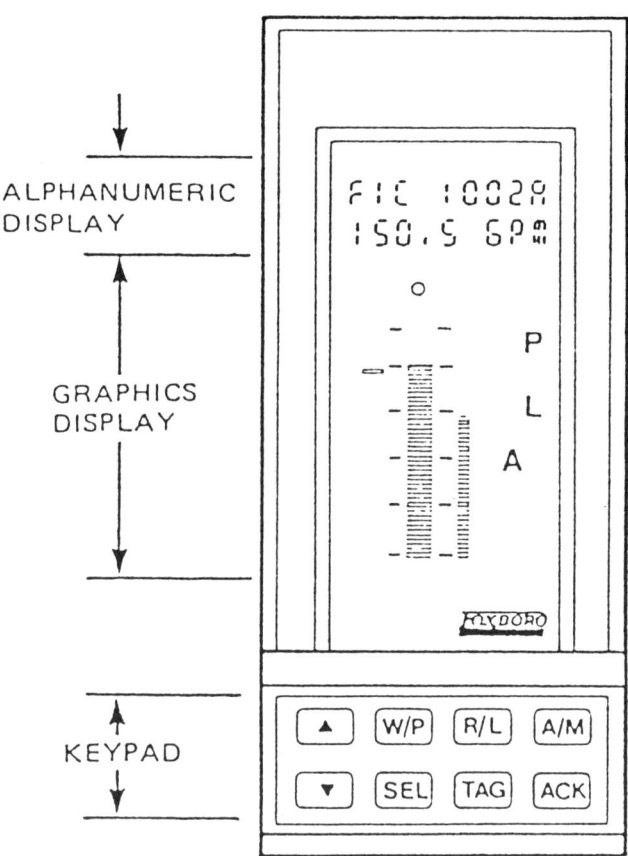

Figure 5.5
The Front Panel of the Foxboro EXACT™ Controller

DFCT   Derivative factor. Allows derivative influence to be reduced or increased by a factor DFCT. Default value is 1. It can be chosen in the interval [0,1]. Setting DFCT = 0 results in a PI controller.

BUMP   Step size in the pretuning phase. The default value is 8%.

Autotuning

Experiences

The Foxboro adaptive controller was announced in October 1984. It has been tested extensively in the field since the spring of 1983. The practical applications experience have been favorable.

5.4 THE SATT CONTROL INSTRUMENTS® AUTOTUNER

The autotuner manufactured by Satt Control Instruments is based on estimation of one point on the Nyquist curve using the method of relay oscillations, as was discussed in Section 5.2, and a modified Ziegler-Nichols method for determining the controller parameters. The ideas behind the autotuner are described in a number of publications, including Åström and Hägglund, 1983, 1984a-c, and Hägglund and Åström, 1985a. It is also patented (see Hägglund and Åström, 1985b).

The autotuner comes in two different versions. One version is a small (about 45 loops) DDC system (Direct Digital Control system) for industrial process control called SDM-20™. The tuner can be connected to tune any loop in the system. The other version is a single-loop controller called ECA-40™ where the normal operating modes manual and automatic have been augmented with a third, a tuning mode. The ECA-40 is shown in Figure 5.6. Tuning is activated upon demand from the operator.

Process Modeling

The parameter estimation is made by the relay method discussed in Section 3.3. The relay has a hysteresis that is set automatically, based on a determination of the measurement noise level. The relay amplitude is also adjusted automatically to give a prescribed amplitude of the oscillation. The determination of amplitude and period is based on simple peak and zero-crossing detection.

Tuning is performed in the following way (see Figure 5.2). The process is brought to a desired operating point, either by the operator in manual mode or by a previously tuned controller in automatic mode. When the loop is stationary, the operator presses the tuning button. After a short period, when the noise level is measured automatically, a relay with hysteresis is

Autotuning

Figure 5.6
The Satt Control Instruments Controller ECA-40™

introduced in the loop, and the PID controller is temporarily disconnected (see Figure 5.2). The hysteresis of the relay is determined automatically from the noise level.

The relay with hysteresis makes the system oscillate, as was discussed in Section 3.3. During the oscillation, the relay amplitude is adjusted so that a

Autotuning

desired level of the oscillation amplitude is obtained. When an oscillation with constant amplitude and period is obtained, the relay experiment is interrupted. The location of the point on the Nyquist curve that corresponds to the oscillation frequency can be calculated from the amplitude of the oscillation and the relay characteristics. The PID parameters are then determined, and the PID controller is activated.

Control Design

The controller design is based on a modified Ziegler-Nichols procedure. A test is provided to determine if derivative action is required. It has been found in practice that derivative action is used quite frequently and that it improves the performance. There are, however, exceptions, such as level control, when derivative time should not be used. It has been found that the identification procedure is very effective in detecting such cases.

The autotuner also has gain scheduling. A table is provided in which a variable that characterizes the operating conditions and the corresponding controller parameters are stored. A facility for automatic switching between the different controller parameters is also provided.

Operator Interface

The initial relay amplitude is given a default value that is suitable for most process control applications. This parameter is not critical since it will be adjusted after the first half period to give an admissible amplitude of the limit cycle oscillation. The operation of the autotuner is then very simple. To use the tuner, the process is simply brought to an equilibrium by setting a constant control signal in manual mode. The tuning is activated by pushing the tuning switch. The controller is automatically switched to automatic mode when the tuning is complete.

The width of the hysteresis is set automatically, based on measurement of the noise level in the process. The lower the noise level, the lower the amplitude required from the measured signal. The relay amplitude is controlled so that the oscillation is kept at a minimum level above the noise level.

The following are some optional settings that may be set by the operator:

(1) Control Design [1, 2, 3]. The desired design of a control loop may differ from the default design procedure in the autotuner. By choosing design [1], a faster response is obtained. Design [2] is the default

Autotuning

design, and design [3] gives a slower response. The control design can be changed either before or after a tuning.

(2) Reset [Yes/No]. Some information from a tuning is saved and used to improve the accuracy of the following tunings, including the noise level, the initial relay amplitude, and the period of the oscillation. If a major change is made in the control loop, e.g., if the controller is moved to another loop, the operator can make a reset of the tuning information.

(3) Initial relay amplitude. In some very sensitive loops, the initial input step of the relay experiment may be too large. The initial step can then be decided by the operator.

(4) Gain scheduling reference. The switches between the different sets of PID parameters in the gain-scheduling table can be performed with different signals as the gain scheduling reference. The scheduling can be based on the input signal, the output signal, or another external signal. The gain scheduling is active only when a reference signal is configured.

A Practical Example

The properties of the autotuner are illustrated in Figure 5.7. The example is taken from a paper mill where the level in a pulp vessel is controlled. Originally, only a P controller was used, i.e., there was neither integral action nor derivative action. This gave rise to a steady-state error between the desired level and the actual level, as shown in Figure 5.7. The tuning took about 2 minutes and resulted in a PI controller. This is an example of how the autotuner can decide that the derivative term is not appropriate for certain processes. Inclusion of the integral term in the control function removes the earlier permanent level error. The response of the system to step changes in the set point also shows that the parameters selected for the controller are appropriate.

Experience

The autotuner was announced in May of 1984. Before that it was tested extensively in the laboratory and in the field. It has proven particularly useful in commissioning of new industrial plants where there is no prior knowledge of controller parameters. It has been shown that commissioning

Autotuning

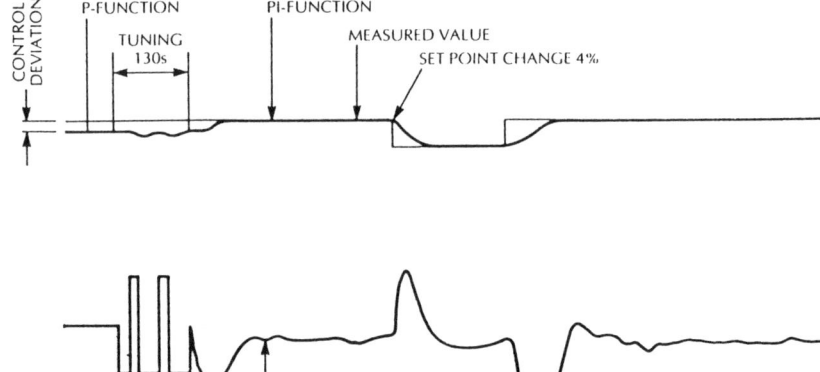

*Figure 5.7
Results Obtained by the Satt Control
Instruments Autotuner Applied to Level Control*

time can be shortened considerably by using the autotuner, especially for plants that have many slow loops. Simplicity is the major advantage of the autotuner. It is very easy for the operator to use. This has proven particularly useful for plants that do not have qualified instrument engineers and for operation during the night shift when instrument engineers are not available. It is also easy to explain the autotuner to the instrument engineers. This has contributed to its quick acceptance.

5.5 THE LEEDS & NORTHRUP ELECTROMAX V™

This is a single-loop controller primarily intended for temperature control. It is an enhancement of a conventional well-proven digital PID controller. The tuning is based on estimation of the parameters of a second-order discrete time model. The PID parameters are computed from this model. The PID controller uses fast sampling, so it is essentially equivalent to a continuous time controller. Slower sampling is used in the parameter estimation.

Autotuning

The controller can operate in three different modes: fixed, self-tune, and self-adaptive. In the fixed mode, it is an ordinary fixed-gain PID controller. In the self-tune mode, a perturbation signal is automatically introduced, a model of the process is estimated, and the PID parameters are computed from the model. The parameters are displayed to an operator who may accept or reject the new parameters. In the self-adaptive mode, the parameters are updated continuously.

The desired closed-loop response time T_r (given as the time to reach 90% of a commanded step response) is the key specification, which must be given by the operator. This specification is critical.

Many aspects of the controller are described in the open literature (see Hoopes et al., 1983, and Yarber, 1984). However, few details are proprietory, and some ideas are also patented.

Process Modeling

Parameter estimation is performed in the self-tune and the self-adaptive modes. In both cases the estimation is performed in closed loop. A sequence of set point changes is generated automatically to ensure that the estimation is based on good data. The changes are cycles of positive and negative pulses, as shown in Figure 5.8. The pulse height (acceptable set point upset) is set by the operator. Some guidelines for the choice are given. It should be larger than 1% of the signal span but not so large that it drives the system into saturations. The value should thus be smaller than half the proportional

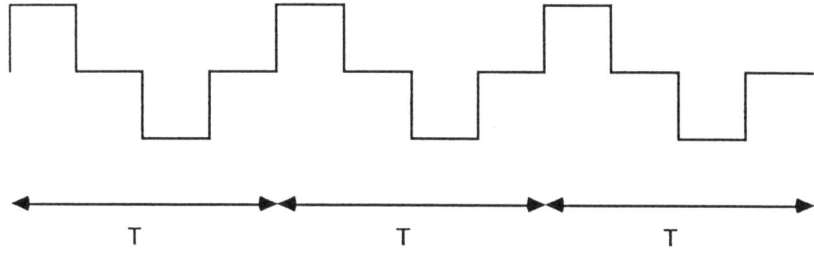

Figure 5.8
Set Point Changes Used in the Identification Mode

Autotuning

band. The cycle time is computed from the response time, which is also set by the operator. The goodness of fit is determined after each cycle. When the fit is good enough, a message is given. The operator may view the controller parameters obtained. If the fit is still poor after five cycles, the procedure is aborted and a message is given.

The parameters a_1, a_2, b_1, and b_2 in the discrete time model

$$y(t) + a_1 y(t-h) + a_2 y(t-2h) = b_1 u(t-h) + b_2 u(t-2h)$$

are estimated recursively by the instrumental variable method. The following formulas are used:

$$\theta(t+1) = \theta(t) + \frac{P(t)}{1 + \psi^T(t)P(t)\varphi(t)} \psi(t)[y(t+1) - \theta^T(t)\varphi(t)]$$

$$P(t+1) = P(t) - P(t)\psi(t) \frac{1}{1 + \psi^T(t)P(t)\varphi(t)} \varphi^T(t)P(t) + D$$

where D is a positive matrix, θ is a vector of parameter estimates

$$\theta = [a_1 \ a_2 \ b_1 \ b_2]^T$$

and φ is a regression vector

$$\varphi(t) = [-y(t) \ -y(t-1) \ u(t) \ u(t-1)]^T$$

Furthermore, ψ denotes the vector

$$\psi(t) = [-y_m(t) \ -y_m(t-1) \ u(t) \ u(t-1)]^T$$

where y_m is the model output computed from

$$y_m(t) = \theta^T(t-1)\psi(t-1)$$

The sampling period used in the parameter estimation scheme should be larger than the time delay of the system. No details are given, but it seems reasonable to relate it to the specified response time.

All input and output signals are high pass filtered to remove the levels from the signals before they enter into the estimation algorithm. A digital filter with the pulse transfer function

$$H(z) = \frac{b(z-1)}{z-a}$$

is used. The selection of the filter constants has not been published, but it seems reasonable to choose the filter parameters based on the specified response time.

Some safeguards are also introduced. The estimation is interrupted when the changes in the control signal are small or when the control error is below a certain limit (see Yarber, 1984).

Control Design

The control design is a multi-step procedure. It is desired to obtain a given response time to command signals. This is achieved by specifying the closed-loop transfer function

$$H_m(z) = \frac{1-a}{z-a}$$

where $a = \exp(-2.3\, h/T_r)$, h is the sampling period used in the system identification, and T_r is the specified 90% response time. A simple control design that realizes the desired closed-loop dynamics is obtained by a discrete time controller with the transfer function

$$H_r(z) = \frac{H_m(z)}{H_p(z)(1 - H_m(z))} = \frac{1-a}{b_1} \frac{z^2 + a_1 z + a_2}{(z-1)(z + b_2/b_1)}$$

This has the same structure as a PID controller. Notice that the process poles are canceled by the zeros of the controller. There are also some difficulties. To have a conventional PID controller, the parameters b_1 and b_2 should have different signs. The parameter estimates will not necessarily have this property. The discrete time controller given above is not used. Instead, the parameters of a continuous time controller are computed from the above expression using a proprietory method (see Anderson and Arcara, 1983). These parameters are then entered in a conventional digital PID algorithm with fast sampling (0.5 seconds).

Operator Interface

Figure 5.9 shows the front panel of the controller. There is a door on the panel to hide some touch switches. When the door is closed, the normal auto/manual switch is accessible as well as switches to increase and decrease the set point of the control variable. Additional switches to change the controller parameters and to interact with the tuning functions are behind the door. No computer terminal is required. The man-machine interface is

Autotuning

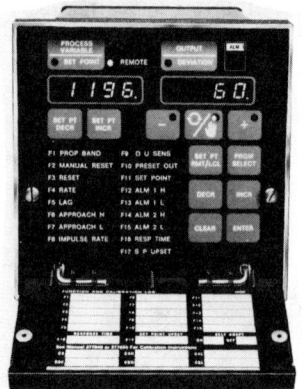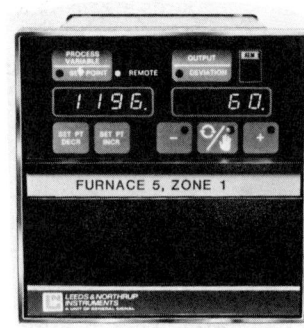

Figure 5.9
The Front Panel of the Electromax V™ Controller with the Door Open (right) and Closed (left)

based on a program that cycles through different functions, which are displayed. The program is activated by pushing the "program select" switch. A particular function is selected by releasing the switch when the selected function is displayed. The parameters can then be modified and additional functions can also be selected.

The following variables and functions are accessible:

F1, Proportional band $(100/k)$
F2, Manual reset
F3, Reset $(60/T_i)$
F4, Rate $(60T_d)$
F5, Lag. Filtering of measured value.
F6 and F7, Anti-windup functions
F8, Impulse rate
F9, $D\ U$ Sense
F10, Preset out
F11, Set point
F12–F15, Alarm limits
F16, Response time
F17, Set point upset
AF1, Computed controller gain $(100/k)$

AF3, Computed reset ($60/T_i$)
AF4, Computed rate ($60T_d$)
P1, Process variable (y)
P2, Deviation
P3, Output (u)

The parameters associated with the tuning are F16 (which gives the desired response time, T_r) and F17 (which gives the amplitude of the admissible set point changes used in identification). The controller parameters obtained by the tuning procedure are AF1, AF3, and AF4. When these parameters are displayed, the adaptive functions Identification, Examine, Fixed, or Self-adaptive can be selected.

Prior Information and Pretuning

To use the controller, it is necessary to specify five numbers: nominal values of the PID parameters, the process response time, and the admissible set point upset. The nominal values of the controller parameters are needed because the estimation is done in closed loop. The process response time (T_r) is defined as the time it takes for the open-loop step response to reach 90 % of the steady-state value. This number is used to determine the period of the perturbation signal in the identification phase, the sampling period of the discrete time model, and the desired response of the closed-loop system. The admissible set point upset gives the amplitude of the pulses used in the identification phase.

The controller parameters must be specified so that a reasonable closed-loop performance is obtained. The desired response time is critical: it is safe to choose a large value; too small a value may lead to instability.

A manual pretuning procedure is suggested to obtain the five parameters necessary to initiate the adaptation. The procedure is a modified Ziegler-Nichols procedure. An open-loop step response is first carried out to determine the open-loop process gain (k_p). The controller is then closed with proportional control only. The gain is chosen as $3/k_p$. The closed-loop step response is then measured. The process time delay (L) and the 90% response time (T_1) are then determined from the step response. The initial controller settings are $k = 3/k_p$, $T_i = T_1$, and $T_d = L/2$. The allowable set point change is chosen as $15\,k_p$. The response time of the controller is chosen as T_1. This choice is a reasonable guarantee that the specifications can be realized.

Autotuning

Commissioning

A typical commissioning proceeds as follows. The controller is set to manual, and the process variable is brought close to the desired operating point under manual control. The manual pretuning procedure is carried out to obtain nominal parameter values. The identification mode is then activated, and improved controller parameters are computed. The parameters can be inspected. They may also be introduced into the control algorithm on operator demand. In the fixed mode the controller runs like an ordinary fixed-gain PID controller. In the adaptive mode they are continuously updated, and the operator can demand to repeat the identification if he is not satisfied with the controller parameters.

Experiences

The Electromax V self-tuner was announced in August of 1981, and almost 20,000 have been installed. About 20% of these have the self-tuning option. Some experiences of their use are summarized in Yarber (1984). The majority of applications are in temperature control. The experiences are generally quite favorable, although it is noted that adaptive control is not a panacea for everything. Most of the benefits are derived from self-tuning, although there are a number of cases where the continuous adaptation has proven to be very profitable. Difficulties in using the controller have been observed with processes that have unsymmetric process response (typically, heating and cooling), very rapid parameter variations, or very strong nonlinearities. The controller cannot be applied to processes, such as silicon crystal growing, that do not tolerate the process upsets required in the identification phase. Difficulties with controllers used in the self-adaptive mode have also been found under operating conditions where the measured value is suddenly disconnected. The parameter estimation is then done on totally irrelevant data. The remedy is to stop the parameter updating when the output is disconnected.

5.6 THE TURNBULL CONTROL SYSTEMS 6355™ AUTOTUNING CONTROLLER

A single-loop autotuner developed by Turnbull Control Systems in the UK is based on a well-proven PID control algorithm that has been widely used for a long time. The PID controller uses fast sampling (0.04s), so it is essentially equivalent to an analog controller. The process modeling is based on fitting a second-order discrete time model. This model is then converted to a continuous time transfer function. The control design is a frequency domain method where the key specification is a phase margin of 60°. There are several interesting features in the man-machine interface.

Process Modeling

The process modeling is based on estimation of the parameters of a second-order discrete transfer function (see Section 3.4). The transfer function fitted is of the form

$$H(z) = \frac{(z+1)(b_0 z + b_1)}{z^2 + a_1 z + a_2} z^{-d}$$

The reason for choosing this particular form is that the controller is sampled so fast that the operation is essentially continuous time. The signals are filtered and the mean values are removed before they are introduced into the estimator. The sampling period used for the parameter estimation is crucial. It is chosen approximately as one tenth of the dominant closed-loop time constant. The parameter estimation is done by recursive least squares with a forgetting factor close to one. The estimation is carried out in closed loop. Several safeguards are introduced. The estimation is discontinued if the change in the process output is too small. A perturbation signal can also be introduced to improve the model accuracy. The amplitude of the perturbation signal can be preset by the instrument engineer. A 'confidence factor' for the model (0–100%) is computed based on a normalized and filtered prediction error. This factor also depends on several checks for validity of the process model; e.g., if the model is open-loop stable, and if the model has the correct sign of the steady-state gain. The confidence factor is displayed to help assess the quality of the estimated model. Later versions of the controller also contain a Smith predictor compensator for time delays.

Control Design

The full details of the control design are not made public. The principles are, however, given in the manuals. The discrete time model is first converted to a continuous time transfer function by using the Tustin transformation, i.e.,

$$G(s) = H((1+sT/2)/(1-sT/2))$$

Autotuning

Several types of controllers can be selected (P, PI, PD, or PID). The PID controller used is of a special form. The parameters are selected so that a phase margin of 60° is obtained with an M_p value of 1.12. The same criteria are used also for PI and PD control. The slope of the log gain/phase curve is also evaluated around this point to ensure that pure time delay does not result in a poor gain margin. As a result of this, derivative action may be removed altogether from the recommended PID values, even though the operator initially requested a PID design. The controller uses a sampling period of 0.04s. It has facilities to avoid integrator windup (see Section 2.3). The high frequency derivative gain $N = 4$ is used.

Pretune Mode

The choice of sampling period is crucial for a successful parameter estimation. Since the parameter estimation is performed in closed loop, it is also necessary to have some preliminary values of the controller parameters that

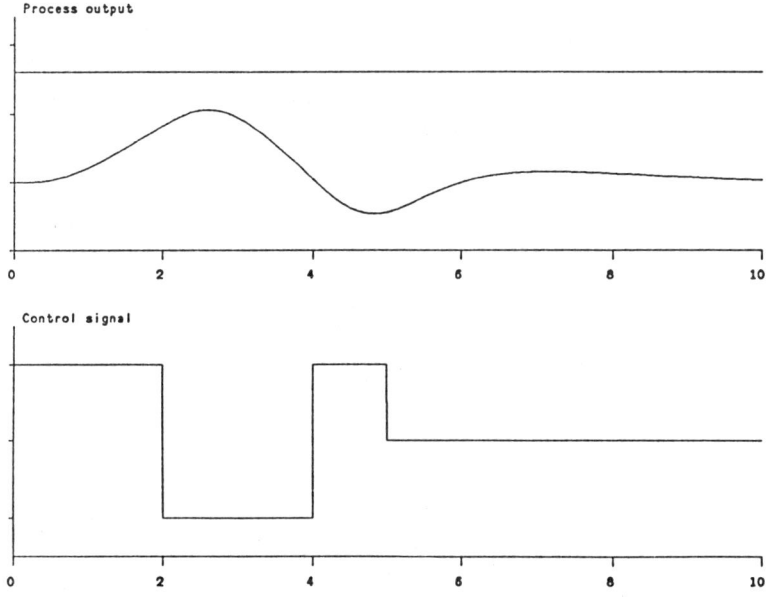

Figure 5.10
Inputs and Outputs during the Pretune Experiment

Autotuning

will give a reasonable control performance. A pretune mode is included to obtain this information. The pretuning is based on a special transient response test. A nominal test amplitude is specified. A step of this magnitude is applied. When the output reaches half of the specified value, the signal is reversed and then switched one more time, as is indicated in Figure 5.10. The time delay of the process, the controller settings, and a proper sampling period are determined from the results of the pretuning experiment.

Operator Interface

A picture of the front panel is shown in Figure 5.11. This looks like a typical single-loop controller with the normal functions for display and switching between manual and automatic mode and facilities to change the set point of the control variable. It thus has two modes: manual and automatic. The parameter estimator runs continuously in the automatic mode. New parameters can be introduced on demand from the operator.

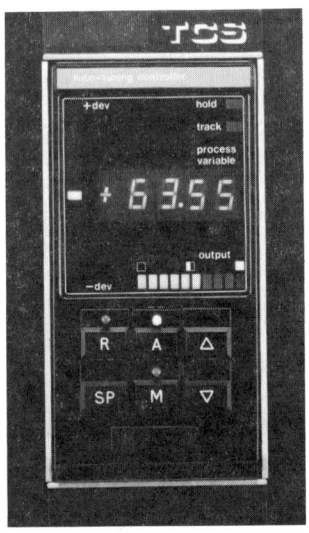

Figure 5.11
The Front Panel of the TCS 6355 Autotuning Controller

Autotuning

To interact with the system, it is possible to connect a hand-held terminal or a more advanced graphics terminal. Access to the following parameters is through the terminal:

IM   Identification Mode. Selects the inputs used in the modeling and the control form P, PI, PD, or PID.

DT   Delay Time. The time delay in the system; automatically set during pretuning but can be altered by the operator.

AD   Auto-Mode Deviation. The amplitude of the allowable set point perturbation.

ID   Initial Deviation. The allowable deviation in process variable during the pretune test.

OD   Output Deviation. The amplitude of the change of the control variable in the pretune mode. This variable must be so large that the process output changes by at least 50% of ID.

T?   This parameter takes three values: 0 for continue tuning, 1 to retune regulator constants in auto-mode, and 2 to execute the pretune mode.

SL   Local set point.

XP, TI, TD Current control parameters.

RP, RI, RD Recommended control parameters.

TT   Sampling time used in recursive parameter estimator.

RT   Recommended sampling time for the recursive parameter estimator.

CF   Confidence factor for the estimated parameters (0-100%).

U?   The recommended parameters are transferred to the controller when this parameter is set to 1. The parameter TT is also automatically set to RT.

Commissioning

A typical commissioning will illustrate the operation of the controller. When applied to an unknown process, the controller is switched to manual and the process is brought to a reasonable operating point manually. The parameters ID and OD are then set and the pretune mode is activated. Nominal values of the controller constant are then obtained and it can be switched to automatic with some caution. The parameter estimation is then

Autotuning

activated, and the current and recommended settings are obtained. An example of the output obtained from a graphical terminal is shown in Figure 5.12. The display also gives the confidence factor and the measured and predicted measurement signals. At this stage, the operator may choose to use the new parameters, or he may choose to wait for a more reliable model. He can also introduce set point changes manually or automatically to get more reliable estimates.

When the controller is running in automatic mode, the estimator is always in operation. If the recommended and the current parameters deviate more than by a user-specified percentage and if the confidence factor is sufficiently high, then the digital readout of the regulator will flash to attract the attention of the operator. The decision to accept the new parameters will, however, always rest with the operator.

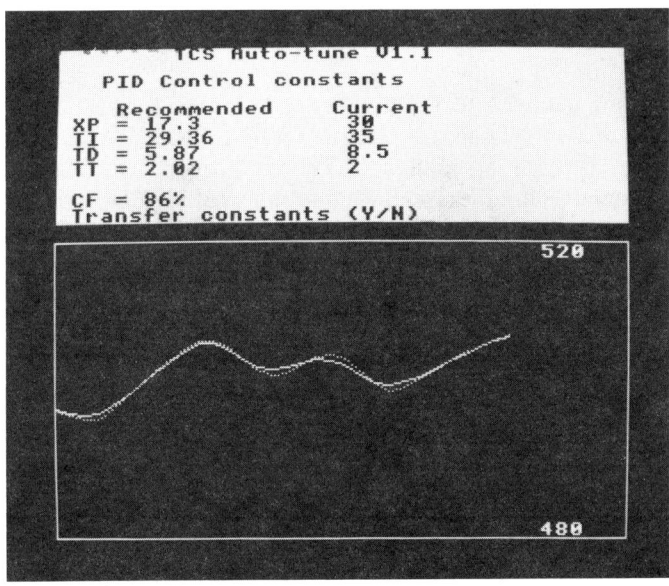

Figure 5.12
Display of Recommended and Current Control Parameters,
Together with Measured Signal and Predicted Measurement Signal

Experience

The TCS 6355 was announced in 1984 and a few thousand systems have been installed. It has been quite favorably received.

5.7 CONCLUSIONS

The technique of autotuning is very new. Even though it has been tried industrially for only a few years, there are currently several thousand loops where autotuning is being used. This is not a negligible number, but it is still a very small fraction of all control loops in operation. From a number of installations with which the authors are familiar, it is clear that autotuning is useful. It can certainly help operators and instrument engineers keep the control loops well tuned. The benefits are even larger for more complex loops. We have noticed, for example, that derivative action is often switched off in manually tuned systems because of tuning problems in spite of the fact that it improves performance. The benefits for automatic tuning are probably even greater for more complicated loops with dead time compensation or feedforward.

Concerning the particular method to use, it is too early to draw definite conclusions. There are many different ways to determine process characteristics, many methods for design of PID controllers, and many ways of combining such techniques to create autotuners. Judging from the systems that are now on the market, it appears that many different ideas have been successfully implemented. However, some patterns do emerge. It appears that more sophisticated methods require more prior information. This is probably what has led to the introduction of the pretune mode, which often has been an afterthought. It would seem that a useful approach to this problem is to combine several different approaches.

Conclusions

6

Adaptive techniques are now finding their way out from the universities to industrial process controllers. It is safe to say that over 10,000 loops are adaptively controlled today. Many more loops are tuned automatically by some autotuning device. Adaptive techniques can be used in different ways, and there are many ways to design an adaptive controller or an automatic tuning device. In this book, attention has been focused on methods for automatic tuning of PID controllers. This is quite a new area—in fact, newer than the general purpose adaptive controllers.

The use of autotuning devices offers several advantages. Since automatic tuning is faster than manual tuning, the commissioning time for installation of new processes can be decreased. In some installations, commissioning time has been reduced by 50% when autotuning was used. PID controllers can be tuned more accurately, which improves the process economy. When the controller is tuned manually, both the process model and the design procedure used are often poor. Trial and error is perhaps the most common procedure; otherwise, a simple Ziegler-Nichols approach or a step response method is used. We have shown that automatic tuning can be based on more advanced identification procedures and design methods. Automatic tuning normally results in a PID controller with derivative action. The derivative part is usually excluded when the controllers are tuned manually. Automatic tuning also means that the tuning is made systematically, even for the simplest control loops. Because of these important advantages, we strongly believe that autotuning will soon be a standard feature in most PID controllers.

Conclusions

We have discussed different approaches to autotuning and the basic principles behind them. The different procedures can be characterized according to the identification and design procedures used. A detailed description of these topics was presented in Chapters 3 and 4. There are two basic types of autotuners, the first based on simple transient or frequency response analysis and simple design procedures of the Ziegler-Nichols type. Several temperature controllers are based on these methods. In Chapter 5, we have described an example in detail, namely the process controller by Satt Control Instruments®. The pretuning modes of the Foxboro EXACT™ controller and the TCS-6355™ from Turnbull Control Systems® are also based on transient response methods. The second type is based on estimation of a transfer function model of the process and a pole placement design algorithm. This category includes the TCS-6355™controller and the Electromax V™ from Leeds & Northrup®.

The autotuning devices available on the market today differ with respect to the amount of prior knowledge needed. For the operator, it is very important to have an autotuner that is easy to operate. The information he must give the autotuner is, therefore, a key issue. Autotuners and adaptive controllers based on methods that require much *a priori* information often have a pretune phase that helps the operator in the choice of such information. Since the PID controller is simple, a tuning device must not result in a controller with parameters that are more difficult to set than the old gain, integral time, and derivative time.

The PID structure has been used ever since pneumatic controllers were introduced. Why stick to this structure now when computers allow other, more general structures? There is no easy answer to this question. One reason is, of course, the historical: it is well-known and there are useful and systematic ways to tune them manually. The simplicity is also an advantage for those loops that do not require any sophisticated controller. But there are also some drawbacks. The simple structure gives the controller limited behavior. Control loops with large dead time or other kinds of complex dynamics are hard to control efficiently with PID controllers. Another argument against the PID controller is that when the tuning is made by the computer, there is no reason to have so few parameters. On the other hand, this argument is true only for those loops where control is left totally to the computer. If the operator wants to have the possibility to switch off the adaptation and make manual adjustments, the controller structure must be simple.

Different kinds of controllers are useful for different kinds of users and control problems. A process engineer with a good knowledge of control would probably prefer a more advanced controller where he can adjust the

Conclusions

control in different ways. For this kind of user, the ASEA Novatune™ controller is suitable. This can be viewed as a tool box, where the operator designs the controller structure himself. For more difficult control problems, the controller must often be tailor-made. Simple PID controllers are then not appropriate.

For the medium high instrument engineers, PID controllers are probably a good choice. They can, however, not control processes with longer time delays and oscillatory modes. It is also useful to add feedforward and gain scheduling.

With the introduction of adaptive technique, we can see the first phase in what can be called management of process knowledge. Instead of having just a manually tuned controller in the control loop, a computer with more sophisticated properties is used. This controller includes knowledge about the process it is controlling. This knowledge is obtained from the operator, but also automatically on line from the control performance using the adaptive technique. We have seen only the beginning of this development. With expert control, the next level of process knowledge management will be made.

References

Anderson, P. P. (1983). "Automatic Identification System for Self-Tuning Process Controller," US pat. of. 4,368,510.

Anderson, P. P. and S. A. Arcara (1983). "Self-Tuning of PID Controller by Conversion of Discrete Time Model Indentification Parameters," US pat. of. 4,407,013.

Arcara, S. A. (1983). "Filter Arrangement for Elimination of Unwanted Bias in a Model Reference Process Control System," US pat. of. 4,385,362.

Åström, K. J. (1979). "Simple Self-Tuners 1," Report TFRT-7184, Dept of Automatic Control, Lund Institute of Technology, Lund, Sweden.

Åström, K. J. (1982). "Ziegler-Nichols Auto-Tuners," Report TFRT-3167, Dept of Automatic Control, Lund Institute of Technology, Lund, Sweden.

Åström, K. J. (1983). "Theory and Applications of Adaptive Control," *Automatica, 19*, 471-486.

Åström, K. J. (1987). "Adaptive Feedback Control," *Proceedings*, IEEE, 75, 185-217.

Åström, K. J., J. Anton and K. E. Årzén (1984). "Expert Control," *Automatica, 22*, 277-286.

References

Åström, K. J. and T. Hägglund (1983). "Automatic Tuning of Simple Regulators for Phase and Amplitude Margins Specifications," *Proceedings* of the IFAC Workshop on Adaptive Systems in Control and Signal Processing, San Francisco.

Åström, K. J. and T. Hägglund (1984a). "Automatic Tuning of Simple Regulators," *Proceedings* of the IFAC 9th World Congress, Budapest, Hungary.

Åström, K. J. and T. Hägglund (1984b). "Automatic Tuning of Simple Regulators with Specifications on Phase and Amplitude Margins," *Automatica, 20,* 645-651.

Åström, K. J. and T. Hägglund (1984c). "A Frequency Domain Approach to Analysis and Design of Simple Feedback Loops," *Proceedings* of the 23rd IEEE Conference on Decision and Control, Las Vegas.

Åström, K. J. and B. Wittenmark (1984). *Computer Controlled Systems,* Prentice Hall, Englewood Cliffs, N.J.

Atherton, D. P. (1975). *Nonlinear Control Engineering—Describing Function Analysis and Design,* Van Nostrand Reinhold, London.

Atherton, D. P. (1982). Limit Cycles in Relay Systems. *Electronic Letters 18,* No 21.

Balchen, J. G. and B. Lie (1986). "An Adaptive Controller Based Upon Continuous Estimation of the Closed Loop Frequency Response," *Proceedings* of the 2nd IFAC Workshop on Adaptive Systems in Control and Signal Processing, Sweden, 31-36.

Bristol, E. H. (1967). "A Simple Adaptive System for Industrial Control," *Instrumentation Technology,* June 1967.

Bristol, E. H. (1970). "Adaptive Control Odyssey," Paper 561-570 ISA Silver Jubilee Conference, Philadelphia.

Bristol, E. H. (1977). "Pattern Recognition: An Alternative to Parameter Identification in Adaptive Control," *Automatica, 13,* 197-202.

Bristol, E. H., G. R. Inaloglu and J. F. Steadman (1970). "Adaptive Process Control by Pattern Recognition," *Instrum. Control Systems,* 101-105.

Deshpande, P. B. and R. H. Ash (1981). *Elements of Computer Process Control with Advanced Control Applications,* ISA, Research Triangle Park, N.C.

References

Dumont, G. A. (1986). "On the Use of Adaptive Control in the Process Industries," in Morari and McAvoy (Eds.), Chemical Process Control-CPCIII, *Proceedings* Third International Conference on Chemical Process Control, Asilomar, CA, Elsevier, Amsterdam.

Elgerd, O. I. and W. C. Stephens (1959). "Effect of Closed-Loop Transfer Function Pole and Zero Locations on the Transient Response of Linear Control Systems," *Applications and Industry*, 42, 121-127.

Gawthrop, P. J. (1982). "Self-Tuning PI and PID Controllers," IEEE Conference on Applications of Adaptive and Multivariable Control, Hull.

Gelb, A. and W. E. Vander Velde (1968). *Multiple-input Describing Functions and Nonlinear Systems Design*, McGraw-Hill, New York.

Gille, J. C., M. J. Pelegrin and P. Decaulne (1959). *Feedback Control Systems*, McGraw-Hill, New York.

Hägglund, T. (1981). "A PID Tuner Based on Phase Margin Specifications," Report TFRT-7224, Dept of Automatic Control, Lund Institute of Technology, Lund, Sweden.

Hägglund, T. and K. J. Åström (1985a). "Automatic Tuning of PID Controllers Based on Dominant Pole Design," *Proceedings* of the IFAC Conference on Adaptive Control of Chemical Processes, Frankfurt.

Hägglund, T. and K. J. Åström (1985b). "Method and an Apparatus in Tuning a PID-Regulator," U. S. Patent Number 4549123.

Hang, C. C., C. C. Lim and S. H. Soon (1986). "A New PID Auto-Tuner Design Based on Correlation Technique," *Proceedings* on the 2nd Multinational Instrumentation Conference, China.

Hawk, Jr. W. M. (1983). "A Self-Tuning, Self-Contained PID Controller," *Proceedings* ACC-1983, 838-842.

Hess, P. F. Radke and R. Schumann (1987). "Industrial Applications of a PID Self-Tuner Used for System Start-up," *Preprints* of the IFAC World Congress, 3 21-26.

Higham, E. H. (1985). "A Self-Tuning Controller Based on Expert Systems and Artificial Intelligence," *Proceedings* of Control 85, England, 110-115.

Hoopes, H. S., W. M. Hawk Jr. and R. C. Lewis (1983). "A Self-Tuning Controller," *ISA Transactions*, 22, 49-58.

References

Horowitz, I. (1963). *Synthesis of Feedback Systems,* Academic Press, New York.

Kraus, T. W. and T. J. Myron (1984). "Self-Tuning PID Controller Uses Pattern Recognition Approach," *Control Engineering,* June 1984, 106-111.

Kurz, H. (1979). "Digital Parameter-Adaptive Control of Processes with Unknown Constant or Timevarying Dead Time," *Preprints* of the 5th IFAC Symposium on Identification and Parameter Estimation, Darmstadt.

Landau, I. D. (1979). *Adaptive Control—The Model Reference Approach,* Marcel Dekker, New York.

Leeds and Northrup (1983). "Automatic Identification System for Self-Tuning Process Controller," US pat. of. 4,368,510.

Leeds and Northrup (1983). "Filter Arrangement for Elimination of Unwanted Bias in a Model Reference Process Control System," US pat. of. 4,385,362.

Leeds and Northrup (1983). "Self-Tuning of PID Controller by Conversion of Discrete Time Model Identification Parameters," US pat. of. 4,407,013.

Lukas, M. P. (1986). *Distributed Process Control Systems—Their Evaluation and Design,* Van Nostrand Reinhold, New York.

Martin-Sanchez, J. M. and G. A. Dumont (1985). "Industrial Comparison of an Auto-Tuned PID Regulator and an Adaptive Predictive Control System (APCS)," *Proceedings* of the IFAC Workshop on Adaptive Control of Chemical Processes, Frankfurt/Main, Pergamon Press.

McMillan, G. K. (1983). *Tuning and Control Loop Performance,* ISA, Research Triangle Park, NC.

McMillan, G. K. (1986). "Advanced Control Algorithms: Beware of False Prophecies," *InTech,* January, 55-57.

Morris, H. M. (1987). "How Adaptive are Adaptive Process Controllers," *Control Engineering 34-3,* 96-100.

Mulligan, L. H. (1949). "The Effect of Pole and Zero Locations on the Transient Response of Linear Dynamic Systems," *Proceedings,* Institute of Radio Engineers, *37,* 516-529.

References

Nachtigal, C. L. (1986). "Adaptive Controller Performance Evaluation: Foxboro EXACT and ASEA Novatune," *Proceedings* ACC-86, 1428-1433.

Nachtigal, C. L. (1986). "Adaptive Controller Simulated Process Results: Foxboro EXACT and ASEA Novatune," *Proc.* ACC-86, 1434-1439.

Nishikawa Y. N. Sannomiya, T. Ohta and H. Tanaka (1984). "A Method for Auto-Tuning of PID Control Parameters," *Automatica*, 20, 321-332.

Radke, F. and R. Isermann (1987). "A Parameter-Adaptive PID Controller with Stepwise Parameter Optimization," *Automatica*, 23, 449-457.

Rivera, D. E., M. Morari and S. Skogestad (1986). "Internal Model Control—4. PID Controller Design," Ind. Eng. Chem. *Proceedings* Des. Dev, 25, 252-265.

Shinskey, F. G. (1979). *Process-Control Systems Application/Design/Adjustment*, Second Edition, McGraw-Hill, New York.

Smith, C. L. (1972). *Digital Computer Process Control*, Intext Educational Publishers, Scranton, P. A.

Truxal, J. (1955). *Automatic Feedback Control System Synthesis,* McGraw-Hill, New York.

Tsypkin, J. A. (1958). *Theorie der relais systeme der automatischen regelung.* R. Oldenburg, Munich.

Wittenmark, B. and K. J. Åström (1980). "Simple self-tuning controllers," in Unbehauen, H (editor), *Methods and Applications in Adaptive Control*, Springer, Berlin.

Yarber, W. H. (1984). "Electromax V Plus, A logical Progression," *Proceedings*, Control Expo '84.

Zervos, C., P. R. Belanger and G. A. Dumont (1985). "On PID Controller Tuning Using Orthonormal Series Identification," *Preprints* of the IFAC Workshop on Adaptive Control of Chemical Processes, Frankfurt/-Main, 13-18.

Ziegler, J. G. and N. B. Nichols (1942). "Optimum Settings for Automatic Controllers," *Trans. ASME*, 64, 759-768.

About The Authors

Karl Johan Åström

Dr. Karl Johan Åström received an MS in engineering physics and a PhD in mathematics and control from the Royal Institute of Technology in Stockholm. After teaching there following his graduation and working on inertial guidance for the Research Institute of National Defense in Stockholm, Dr. Åström joined the IBM Nordic Laboratory to work on theory and applications of computerized process control. In 1965 he was appointed professor to the chair of automatic control at Lund Institute of Technology/University of Lund. He has also held visiting professorships at Brown University, USC, University of Chicago, Washington U., U. of Texas, Imperial College of London, University of Manchester, and University of Warwick.

An associate editor of *Automatica, The International Journal on Control, Journal of Mathematical Analysis and Applications, Time Series Analysis, Mathematical Biosciences,* and *Information Sciences,* he is a consulting editor of *IEEE Transactions in Automatic Control* and has been on the editorial board of *Journal of Physics E., Scientific Instruments.*

Dr. Åström is a member of the Royal Swedish Academy of Sciences, the Swedish Academy of Engineering Sciences, the Royal Physiographical Society, the Scientific Council of the Swedish Board of Technical Development, and was vice chairman of the Theory Committee of IFAC, where he has been a member of the IFAC Council from 1981 to the present.

A Fellow of IEEE, Dr. Åström's other honors include the Callender Silver Medal (IMC, London), the Wilhelm Westrup Prize (Royal Physiographical Society, Lund), the Rufus Oldenburger Medal (ASME), the Great Prize (Royal Institute of Technology), the Chester Carlsson Medal (IVA), and the Quazza Medal (IFAC).

Tore Hägglund

Dr. Tore Hägglund was born in Lund, Sweden in 1954. He received an MS degree in engineering physics in 1978 and a DSc degree in automatic control in 1984, both from the Lund Institute of Technology. His main research interests are in the areas of adaptive control, autotuning, and PID control. In 1985 he joined the R&D department at Satt Control Instruments, where he is working with design and implementation of autotuners, adaptive control, and other advanced control systems. Dr. Hägglund has written several papers and holds two patents.